Astrophysics and Space Science Library

For other titles published in this series, go to
http://www.springer.com/series/5664

E.F. Milone • C. Sterken
Editors

Astronomical Photometry

Past, Present, and Future

Springer

Editors
E.F. Milone
Department of Physics and Astronomy
University of Calgary
Calgary, AB, T3A 2J5, Canada
milone@ucalgary.ca

C. Sterken
Vrije Universiteit Brussel
Pleinlaan 2, Brussels, Belgium
csterken@vub.ac.be

ISSN 0067-0057
ISBN 978-1-4419-8049-6 e-ISBN 978-1-4419-8050-2
DOI 10.1007/978-1-4419-8050-2
Springer New York Dordrecht Heidelberg London

Library of Congress Control Number: 2011921729

Printed on acid-free paper

Springer is part of Springer Science+Business Media (www.springer.com)

Preface

No astrophysical theory can be tested without data, and those that deal with predictions of visible objects in the universe often require observational data. The precise and accurate measurement of electromagnetic data is called photometry. In this volume we discuss from both physical and historical perspectives, the elements and practice of astronomical photometry applied to the electromagnetic spectrum from the near ultraviolet to the middle infrared, roughly between 200 and 20,000 nm or 0.2 and 20 μm.

The history of astronomical precision begins with the ancient Greeks, among whom Hipparcos (~190 – ~120 B.C.) provided the first quantitative measurements of stellar "magnitudes" in a catalogue. Photometric precision progressed very slowly until the development of the telescope and the first measures of comparative brightness of the Sun and Moon. Only with the end of the nineteenth century did the precision of astronomical visual photometry reach the 2% level, although not frequently. The application of photography provided a greater degree of objectivity to detections, but brightness measurements from photographic plates were still relatively subjective until the development of measuring engines at the beginning of the twentieth century. Even so, the lack of uniformity of the plates' glass and emulsions, coupled with atmospheric effects, conspired to prevent breakthroughs to greater precision. Photoelectric photometry did achieve greater precision, but again, only in the teeth of intrinsic difficulties. CCD photometry, starting in the 1980s gradually became dominant as CCDs became the detectors of choice at most observatories, but there, too, many problems that plagued the photographic plate era returned, with the additional difficulty of the need to calibrate the spectral or passband sensitivities of what have become ensembles of millions of individual detectors.

The highest precision with which an astronomical brightness measurement can be made is 0.0001 magnitude currently, about 0.01% of the value of the measurement. In practice, such precision is difficult to achieve.

The historical developments are outlined and the methods of achieving the highest possible precision in each era are discussed, along with their limitations. A balance is kept between discussions of hardware and software, between techniques and achievements, and between the science of detection and measurement and the astrophysics for which the photometry is carried out.

In the course of this exposition, we discuss both "absolute" and "relative" photometry, the techniques for doing precise photometry under less than pristine skies, and the techniques to provide the best possible results in cases where the skies are indeed "photometric." References are made to calibrations for both ground- and space-based surveys, although we do not discuss in this volume the important topic of astronomical surveys per se, which deserves its own extensive treatment. There are treatments also of the ever important techniques of spectrophotometry and polarimetry, and, in all the fields of astronomical photometry, the promise of further improvements is explored.

This volume on the past, present, and future of photometry combines the views of past and present and perhaps future members and officers of the International Astronomical Union's Commission 25 on Photometry and Polarimetry. The opportunity to combine these views came about through sessions convened at a Historical Astronomy Division meeting held simultaneously with its parent organization, the American Astronomical Society, in Long Beach, California in January, 2009. Almost all of the authors who contribute here presented their views at that meeting, but the present papers are far more than a recapitulation of those necessarily brief presentations. The present writings are expansive and have been made as extensive as their authors required to present full exposition.

The authors of the several papers presented here have related the history of photometry as it leads to their particular work. Thus: at the outset, Sterken, Milone and Young describe the development of photometry from the ancient Greeks down to the CCD age and follow the schools of thought that have led to passband systems to classify stars, account for interstellar reddening, and find their fundamental properties. The papers by Milone and Pel, Ambruster et al., and by Howell, emphasize the high precision attainable by differential photometry, primarily. The Milone and Pel paper considers the advances from visual to photoelectric photometry of individual stars, discussing many of the instruments and techniques that led to these advances, and highlighting the contributions and legacy of Theodore Walraven. The Ambruster et al. paper builds further on that discussion to describe the independent development of a highly successful pulse-counting differential photometer. Howell's work emphasizes the very sharp rise of the precision of CCD photometry. Several papers discuss the challenges of CCD photometry and suggest ways to improve its accuracy. Landolt demonstrates how the work of the twentieth century pioneers of *UBV* photometry such as Harold Johnson has led to the precise and accurate broadband system that can be labeled the Landolt *UBV*(*RI*) system. Wing demonstrates how his 8-color system developed from the early attempts by Johnson, Morgan, and Keenan to classify stars. In the course of doing this, he is able to make a fine distinction between photometric systems that provide a true spectral classification and those that do not. Milone and Young show how the careful consideration of passbands can improve on the legacy of Johnson in the near and intermediate infrared and can lead to very high precision as well as accuracy. Adelman looks at the varying successes of spectrophotometry of the past and how progress is being made to increase the accuracy as well as the precision through new instruments. Cohen discusses all-sky calibration programs, extending from the UV through the intermediate infrared. Both authors discuss the difficulties in the use of Vega as a calibrator, now known to be a flawed standard. The paper by Bastien describes the equally lengthy history of what may be the most precise form of photometry, polarimetry. There is, inevitably, some repetition of the historical developments from paper to paper, but the variety of lessons learned from those separate discussions demonstrates how examining many particular points of view can lead us to a well-rounded appreciation of the development of our science, and of the future that we can envisage for it.

The editors would like to thank the officers of the AAS Historical Astronomy Division in 2008–2009 who supported the Long Beach sessions, and the astronomy editors at Springer, especially Harry Blom and Chris Coughlin, for their encouragement and support.

Calgary, Alberta, Canada
and Brussels, Belgium
June 2010

Eugene F. Milone
Christiaan Sterken

Contents

Contributors

Saul J. Adelman Department of Physics, The Citadel, 171 Moultrie Street, Charleston, SC 29409, USA, adelmans@citadel.edu

Carol W. Ambruster Department of Astronomy and Astrophysics, Villanova University, Villanova, PA 19085, USA, carol.ambruster@villanova.edu

Pierre Bastien Département de physique et Centre de recherche en astrophysique du Québec, Université de Montréal, C.P. 6128, Succ. Centre-Ville, Montréal, QC H3C 3J7, Canada, bastien@astro.umontreal.ca

Martin Cohen University of California, Berkeley, CA 94709, USA, mcohen@astro.berkeley.edu

Steve B. Howell NOAO, 950 N. Cherry Ave. Tucson, AZ 85726, USA, howell@noao.edu

Anthony B. Hull Department of Physics and Astronomy, University of New Mexico, Albuquerque, NM, USA, tony.hull@L-3com.com

Robert H. Koch* Department of Physics and Astronomy, University of Pennsylvania, Philadelphia, PA 19104, USA

Arlo U. Landolt Department of Physics and Astronomy, Louisiana State University, Baton Rouge, LA 70803, USA, landolt@phys.lsu.edu

Eugene F. Milone RAO, Department of Physics and Astronomy, University of Calgary, 2500 University Drive, NW, Calgary, AB, Canada T2N 1N4, milone@ucalgary.ca

Richard J. Mitchell Gravic, Inc., Suite 100, 301 Lindenwood Drive, Malvern, PA 19355, USA, rmitchell@gravic.com

Jan Willem Pel Kapteyn Astronomical Institute, University of Groningen, Groningen, The Netherlands, pel@astro.rug.nl

Christiaan Sterken Vrije Universiteit Brussel, Pleinlaan 2, B-1050 Brussels, Belgium, csterken@vub.ac.be

* Professor Koch passed away on 11 October 2010, as this book was going to press.

Robert F. Wing Department of Astronomy, Ohio State University, Columbus, OH, USA, wing@astronomy.ohio-state.edu

Andrew T. Young Department of Astronomy, San Diego State University, PA-210, 5500 Campanile Drive, San Diego, CA 92182-1221, USA, aty@mintaka.sdsu.edu

Photometric Precision and Accuracy

Christiaan Sterken, E.F. Milone, and Andrew T. Young

Preamble

Discussing developments of photometric precision and accuracy requires, first of all, a clear understanding of what is meant by *precision* and by *accuracy*, as the two concepts are continually confused:

- *Precision*: how finely a result reproduces (statistical estimate or observable)
- *Accuracy*: how close a result is to the true value

Young (1994) puts it like this:

> *By "precision" is meant the repeatability of a measurement, usually under fixed conditions. On the other hand, "accuracy" means the absence of error, as measured against some external standard, such as a set of standard stars."*

1 Introduction

The development of photometric precision begins with the first estimates of visual astronomy. Accordingly, we begin by recapitulating the history of the search for precise photometric measurements. The review begins with purely naked-eye photometry and extends through the most clever techniques used to increase precision through comparisons with artificial or real comparison stars. We discuss the development of photographic photometry and photoelectric photometry down to the CCD era. We conclude with a discussion of how photometry can be improved still further.

The progress in astronomical photometric precision is well illustrated in Fig. 1, updated from Young (1984), which was based in turn on the superb review of Weaver (1946), upon which this historical review also leans.

C. Sterken (✉)
Vrije Universiteit Brussel, Pleinlaan 2, B-1050 Brussels, Belgium
e-mail: csterken@vub.ac.be

E.F. Milone and C. Sterken (eds.), *Astronomical Photometry: Past, Present, and Future*,
Astrophysics and Space Science Library 373, DOI 10.1007/978-1-4419-8050-2_1,

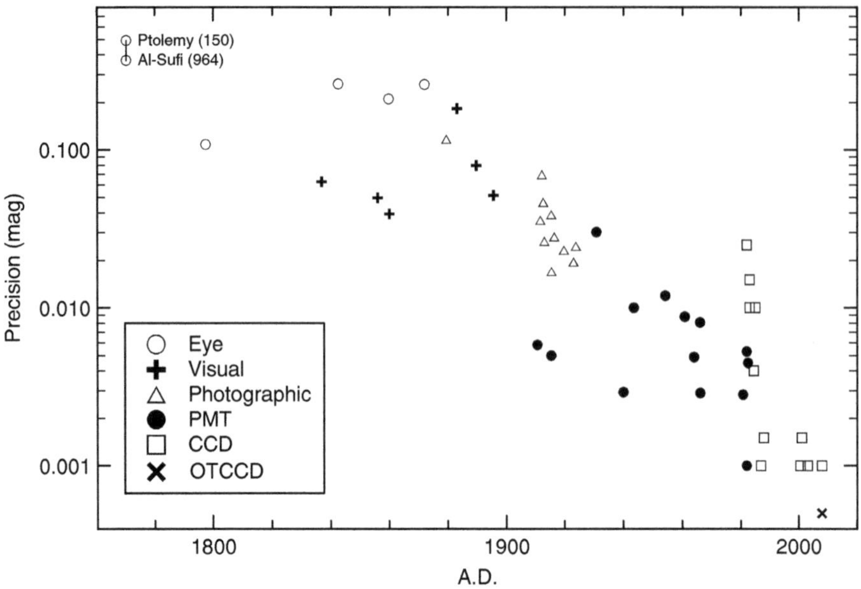

Fig. 1 Evolution of the precision of photometric measurements (adapted from Young (1984)). Eye stands for visual estimates aided by telescope only. Visual refers to all methods where the eye, as a detector, is assisted by attenuating wedges, comparison lamps, etc. The precision estimates for CCDs were taken from Howell's paper (this volume, page 69) and references therein

2 The Visual Era

In this era, which stretches from prehistory to the present, the eye is the primary photometric instrument. The passband, i.e., the wavelength region of the detected spectrum, is thus the broad "visual" sensitivity curve of the, most likely mesopic, retina.[1]

No discussion of photometry can begin without a reference to Hipparcos (2nd c. B.C.), whose magnitude scale was designed to discriminate among the brightnesses of the visible stars. The star catalogue of Ptolemy (2nd c. A.D.; see Toomer (1984), *Almagest*, Books VII, VIII) Toomer (1984) based at least in part on Hipparcos' observations, was the first publication that contained systematic values of stellar brightnesses. The Almagest magnitudes were not intended to be a quantitative numerical scale – indeed, the classes were denoted by letters of

[1] The fact that we can see colors in the first-magnitude stars shows that photopic vision plays an appreciable part about 4 magnitudes above the visual threshold. Most "visual" measurements made with telescopes are probably photopic rather than scotopic – which is why the effective wavelength for the old visual magnitudes is closer to 550 nm than to 500 nm. This transition between photopic and scotopic vision is the reason why the Purkinje effect (see page 6) is significant in visual observations.

the alphabet, though these were often used as digits – but were merely *classes*. Rather than simply classifying stars into two groups, such as "bright" and "faint," a little more detailed classification was used. This is why "class 1" or "first magnitude" (designated α) contains stars of such obviously disparate brightness. Note that these classes were never intended to measure any physical (or even psychophysical) quantity, and that the replacement of the traditional classes by a numerical scale was primarily a nineteenth-century development.

Both William Herschel (1738–1822) and his son, John Herschel (1792–1871) observed stars extensively and made estimates of their relative brightnesses by arranging them in sequences. William Herschel wrote down coded symbols to indicate brightness subintervals. John Herschel extended this work and applied it also to southern skies. Details and references are given in (Hearnshaw 1996, pp. 20–24).

The magnitude scale was reexamined by many astronomers, most notably by Norman Pogson (1856). There were several attempts to generate numbers similar to the traditional magnitude classes before Pogson, such as those by Steinheil (who introduced a logarithmic scale as an "arbitrary assumption") and John Herschel (whose square-root scale was closer to the traditional magnitude numbers); several of these were taken into account by Pogson.

The work by Argelander and the production of the *Bonner Durchmusterung* set the pattern for improvements of precision of absolute photometry. It will be noticed in Fig. 1, however, that prior to the development of photoelectric photometry early in the twentieth century there were two trends in the improvement of precision. This leads us to the topic of differential photometry.

One of the first stellar photometers appears to have been created and used in Sweden by Anders Celsius and Andreas Tulenius in 1740, two years before Celsius described his (inverted) temperature scale to the Swedish Academy of Sciences. It is described in Tulenius's thesis (Tulenius 1740). This was the first of a class of instruments called by Weaver "extinction photometers," that involved a device to dim a star until it disappeared. This type of photometer was subsequently used by the Rev. C. Pritchard in the interval 1881–1885. The extincting element was a wedge of varying optical density inserted in the eyepiece; by measuring the distance moved by the wedge until the star could no longer be seen, a correlate measure of relative star brightness could be obtained. Pritchard, with the help of his associates W. E. Plummer and B. C. Jenkins, used this device to create a catalogue of 2786 stars (Pritchard 1885).

An important step in the advancement of precision was the development of visual stellar photometers, i.e., the application of a technique to diminish the intensity of a light beam by a measurable amount. Examples are the Steinheil, wedge, meridian and Zöllner photometers (see Hearnshaw 2000). In 1858, Karl Friedrich Zöllner introduced a new type of photometer which became a bestseller and which for him led to the first German professorship in astrophysics. His catalogue, containing 425 brightness ratios between bright stars, allows an objective evaluation of the precision of the visual technique; see Fig. 2.

The normal dark-adapted human eye is, in fact, a fine detector, achieving quantum efficiencies as high as 10% under certain circumstances. However, purely visual

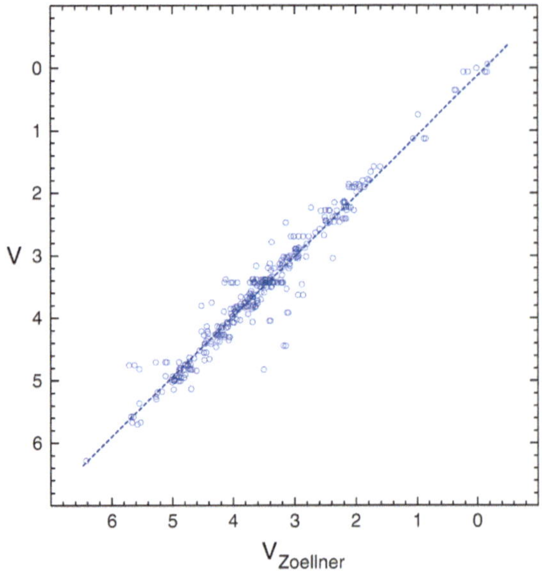

Fig. 2 Zöllner's visual magnitudes compared with *V* from the Bright Star catalogue. The dashed line is a linear fit. Source: Sterken and Staubermann (2000)

photometry is difficult because the memory and recording process are insufficient, and color sensitivity differs slightly from individual to individual. This situation is greatly improved, however, when the eye is used as a comparative instrument. The ability of the human eye to discern small differences in comparative brightness led to the lower (i.e., more precise) trend of the early photometry in Fig. 1, culminating in the Harvard differential photometry of the early twentieth century. Differential photometers are taken up again in the paper by Milone and Pel (2010) in this volume, but we note here that the best differential visual photometry achieved a precision of ~0.02 magnitudes, or ~2%.

In practice, great care was needed to avoid systematic error. In studying such data, it is important to avoid the blind combination of magnitude estimates from different sources. As just one example, Fig. 3 shows the combined photographic-photomultiplier-CCD-visual photometry of Wra 751, a Luminous Blue Variable (LBV) (Sterken et al. 2008). Although the visual estimates mimic the changes of the variable quite closely, they demonstrate significant systematic zero-point deviations – in other words, they have good precision, but very poor accuracy. In particular, the estimates obtained by the two visual observers differ by $0\overset{m}{.}2$ to $0\overset{m}{.}5$ in 2007–2008. These large and variable differences between both data sets result from the combination of the systematic offset between the two observers, and the color effects caused by observing at high airmasses (one observer worked at systematically higher airmasses than the other, including airmasses exceeding 4.0–5.0). Part of the discrepancies between the magnitude sets can be ascribed to the very different effective wavelengths of the "instrumental" bandpasses: the human eye's response in scotopic vision peaks at ~510 nm, whereas the Johnson *V* band peaks

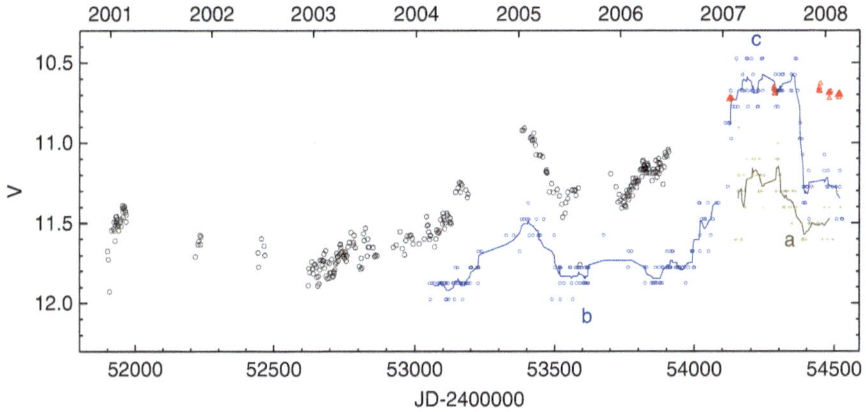

Fig. 3 The evolution of V magnitude during the recent brightening of Wra 751. Visual estimates are represented as running averages a, b and c. The other symbols are photometric measurements. Source: Sterken et al. (2008)

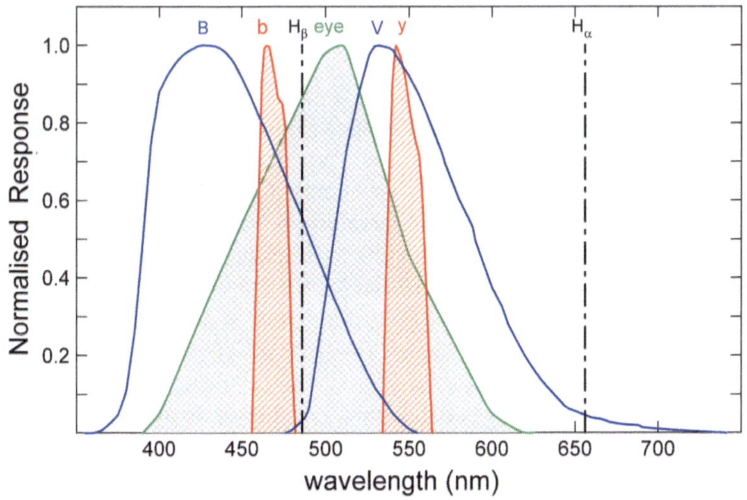

Fig. 4 Normalized photometric response of the B, V, b, and y photometric bands. The response of the human eye in scotopic vision is also shown, as well as the location of the strong and variable Balmer emission lines. Source: Sterken et al. (2008)

near 530 nm.[2] In addition, the eye's response curve includes Hβ and is much wider than V, whereas the V – unlike the Strömgren y – includes Hα in its red tail[3] (see Fig. 4).

---

[2] More importantly, the isophotal wavelength is about 545 nm for V.

[3] The red tail of the Johnson V band is defined by the response of the photomultiplier and not by a filter cutoff.

As both Balmer lines undergo prominent changes during the LBV variability cycles, their effect on the magnitude estimates is hard to quantify. In addition, there are quite a number of FeII and [FeII] emission lines (permitted and forbidden emission lines, respectively) throughout the visual spectrum; see Fig. 1 in Hu et al. (1990). These authors estimate the effect of these emission lines to be $0\overset{\mathrm{m}}{.}03$ in the V passband, although the differences between the y and V photometric scales for LBVs easily may reach $0\overset{\mathrm{m}}{.}2$ (for example, in the case of η Car; see Sterken et al. (1999). This effect has been warned against by Robert F. Garrison (1993):

> *"The small differences in passband definition can be disastrous for unusual objects like supernovae, for which the emission lines develop and recede in addition to moving in radial velocity, so as to move in and out of the boundaries of the pass bands. These differences can lead to differences of 0.4 magnitude or more which is not small for theoreticians."*

Figure 3 also reveals a remarkable discrepancy between the relatively stationary V magnitude of Wra 751 (based on the differential y) and the visual estimates, viz., the $0\overset{\mathrm{m}}{.}5$ drop observed in November 2007, and the $0\overset{\mathrm{m}}{.}2$ decrease as seen later.

3 The Photographic Era

This era begins with the astronomical images captured on Harvard daguerreotype plates, by George P. Bond and his associates. When photography became more generally available, it too was applied to astronomy. Early photographic plates were taken by Rutherfurd at his personal observatory in New York City. Bond resumed his experiments and developed an equation to describe the relation between exposure time and "*photographic power*" (Bond 1859):

$$y^2 = Pt + Q, \tag{1}$$

where t is the exposure time, y is the diameter of the star images, and P and Q are constants.

This collection was eventually given to Columbia University and housed in what became the Rutherfurd Observatory. Eye estimates of brightness from photographic images suffered from a number of difficulties, despite Bond's reasonable argument that photographic magnitude determinations made possible estimates that were less subjective than naked eye estimates: the *Purkinje effect*[4] and color terms, and off-axis effects, among them. As Weaver (1946, p. 288) noted, however, these sorts of problems do not disappear in the combination of telescope and photographic plates.

An eye estimate technique applied to images of variable stars usually involved establishment of a sequence of stars spanning and exceeding the brightness variation of the variable star. Here the differential result depends on the accuracy of

[4] The tendency for the human eye to detect blue objects more readily than red under low light conditions.

the known magnitudes and constancy of the sequence stars as well as the linearity of the relation between magnitude and image diameter, the color terms, and such conditions as the relative locations of the images on the plate.

Photographic accuracy and precision depend on image structure, the stellar spectral energy distribution, and (for eye estimates) the experience of the estimator. The scatter of the points in Fig. 1 indicates typical good results; see Seares et al. (1941), and Seares and Joyner (1943, 1945), nonlinearity of the photographic process, it was much more difficult to establish an accurate Pogson scale photographically than to compare stars of similar brightness; hence, photographic precision was usually much better than its accuracy. After photoelectric techniques solved the scale problem, photography continued to be used to interpolate among stars in crowded fields, especially in star clusters; with long focal lengths (Baum 1962), precision as good as 0.01 or 0.02 magnitude can be reached (e.g., Majewski 1992) if the plates are measured with an iris photometer (Cameron 1951) or microdensitometer. However, 0.1 mag is a more usual value for short-focus cameras.

The large archives of photographic plates preserve the photometric histories of variable stars and other objects, so photographic photometry continues to be done (e.g., Davis et al. (2004) and Schaefer (2010)).

Starting in 1882, following the availability of silver-bromide emulsion in 1871 and subsequent improvements in development techniques, the resulting gains in sensitivity allowed fainter stars to be photographed. This permitted E. C. Pickering (1888) multiple exposures of stars to determine magnitudes. To establish a scale, a series of plates of star trails of the north celestial pole region were taken with the telescope successively stopped down to provide a record of the effects of precisely known ratios of brightness on the trailed images. The estimates were made to the nearest tenth of a magnitude, which provides an indication of the precision that could be achieved with this method. Pickering explored point and extrafocal images as well, and extended the sequence work to equatorial regions and to the Pleiades. Karl Schwarzschild similarly carried out such experiments, as we note below. Pickering also explored the effects of varying exposure time. Reciprocity failure[5], however, limits the value of the latter method.

Plate eye estimates were and still are used in photographic surveys, beginning with the plate measuring photometer developed independently by Harlan Stetson and by Jan Schilt. Other *astrophotometers* included that of Eichner, and the Askania astrophotometer. In 1910, Pickering suggested the method of sending a beam of radiation through a photographic plate and measuring the transmitted light, as a means to measure the density of the photographic image and hence its brightness. This idea was pursued in both the U. S. and in Europe. Harlan True Stetson began a series of experiments in 1911 at the Wilder Laboratory of Dartmouth College that led ultimately to the construction of an instrument that performed this action (Stetson 1916). The instrument used either a 1-mm, or, for smaller images, a 0.5-mm fixed aperture, a 5-mm diameter bismuth-silver thermocouple as a detector (Coblentz

[5] Failure to achieve the same exposure by varying exposure time or source brightness.

1912), and a galvanometer as a measuring device. Working with a "well-known" relation between magnitude and star image diameter,

$$m = a - b\sqrt{D}, \tag{2}$$

where D is the effective diameter of an opaque disk absorbing the same light as the actual star image, Stetson derived an expression between m and δ, the difference in galvanometer excursion between measurements with and without the star image in the aperture:

$$m = a - \beta\delta^{\frac{1}{4}} \tag{3}$$

An excursion of 1 cm was found to be the equivalent of $0\overset{\text{m}}{.}016$.

In a test of the linearity of the relation between m and δ, Stetson compared Pleiades magnitudes determined by Müller and Kempf (1899) with $\delta^{\frac{1}{4}}$ measurements of plates from Dearborn Observatory; Stetson found a probable error of 0.022.

Stetson (1916) then applied his thermopile instrument to the measurement of Harvard plates of the variable star U Cephei, in the process discovering the variability of the star BD + 81°30 (= HD 6006). The uncertainties in the measurement of U Cephei greatly exceeded the measurement errors, leading Stetson to conclude that variations in atmospheric transparency, emulsion variations across plates, and plate color terms from plate to plate contributed to the large range of errors, $0\overset{\text{m}}{.}01$ to $0\overset{\text{m}}{.}20$. Stetson and Carpenter (1923) discuss improvements to the thermopile housing that decreased the time to reach equilibrium, a more sensitive galvanometer (Leeds and Northrup d'Arsonval type R), and a smaller light source, permitting an improvement in the measurement precision: 5 settings per image could be completed in 2.5 min and result in a probable error of $0\overset{\text{m}}{.}007$. They discuss a plate-testing device that imposed artificial stars on a plate. With this apparatus, they could explore such effects as the curvature of the plate, the concavity of which was suspected as the source of variable emulsion thickness. With improved equipment and careful techniques, Stetson and Carpenter found the probable error for a single star image to be $\approx 0\overset{\text{m}}{.}02$. Interestingly, for extra-focal images they obtained a larger probable error, $\approx 0\overset{\text{m}}{.}035$, a result they attributed to the lack of uniformity in plate sensitivity.

At a slightly later time, Jan Schilt (1922) designed and constructed a photometer while at Göttingen and, still later, Leiden. This instrument too employed a fixed aperture to limit the sky background around each star image. Heinrich Friedrich Siedentopf (1935) of Jena, improved on the Schilt design by introducing a variable iris in the beam, expanding the dynamic range that could be measured. Schilt was involved in what Cameron (1951) claimed to be the construction of the first such plate photometer in the U. S. (Eichner et al. 1947). Following this, L. C. Eichner Instruments commercially built such a variable iris device, after the improved design by Siedentopf (1935). Cameron (1951) discusses measurements taken with this device and the errors of the resulting photometry (see the Milone and Pel paper in this volume). He reported reasonably high measurement precision, but also reported that systematic errors generally degraded the overall magnitude determinations. After the 1960s, automatic plate photometers further diminished the labor required and eventually enabled thousands of stars to be measured and recorded per hour. The

GALAXY instrument in Edinburgh (Stoy 1970), and the Cambridge Automatic Plate-Measuring photometer (Kibblewhite 1971) are prime examples. Subsequently, the Perkin–Elmer PDS microdensitometer provided highly precise scanning of any portion or all of a photographic plate. The Machine Automatique á Mesurer (MAMA) employed more than one photoelectric photometer to measure many areas of the plate at once. The automated instruments provided precise positional as well as photometric data. With the scanning engines, digitized densities of the entire plate, stars, nebulosities, galaxies, and so on, are obtained; hence computer processing of all stellar images is necessary.

As with other techniques, differential photographic photometry provided the highest possible precision for photographic work, as evidenced by the work of Jordan (1923, 1929) system RW Com, by Plaut (1940) on WW Draconis = Σ2092A and by Wesselink (1941) on SZ Cam, the eclipsing binary component of the visual binary Σ485.

Among the techniques that were employed to improve photographic precision by these and other astronomers was the use of extra–focal images. This technique involves the distribution of the stellar flux over a greater number of photographic grains than applies for sharp images from an aberration–free, and well-focused telescope. This work was extended to create square star images, (but usually of in-focus stars) a technique that clearly foreshadowed the use of "square star" images in Orthogonal Transfer CCDs.

One device that accomplished this was known as a *Schraffierkassette*, (Meyermann and Schwarzschild 1905) translated in English as a "jiggle camera" or "jiggle–plate." The device rapidly translated the photographic plate in both x and y coordinates. They produced a uniform square exposure image of each star, thus incorporating many silver grains and providing for measurement by a plate photometer square aperture. The benefit of this device over extrafocal images produced by varying the distance of the objective from the focal plane is that effects of lens aberrations, especially color aberration, were avoided. Experiments with extra-focal images were again undertaken by Parkhurst and Jordan (1907) who employed measurements of spot sensitometry on plates of the same emulsion batch as those used for the observations, thus establishing a magnitude-image density calibration. Other techniques of absolute calibrations were devised by Kapteyn, Schwarzschild, and Hertzsprung (these are discussed at length by Weaver (1946, p. 295ff). Hertzsprung's idea was to use an objective grating, the technique successfully used by Plaut and Wesselink in their theses on WW Dra and SZ Cam, respectively. Further details of that work are decribed by Milone and Pel in this volume.

The $\sim 0.^{m}02$ precision attained with the most refined visual differential techniques was achieved even into the mid-twentieth century with the Princeton polarizing photometer, a direct descendant of the Pickering photometer developed at Harvard, and seems to have been at least matched in the era dominated by photographic measurements. In common practice, however, this level of precision was not achieved generally, in either era.

4 The Photoelectric Era

4.1 The Rise of Photoelectric Photometry

The origins of photoelectric photometry (frequently given the acronym PEP) have been discussed at length by Weaver (1946) and by Hearnshaw (1996) and briefly by Butler and Elliott (1993), so we give only a very basic summary here. The pioneering names are William H. S. Monck (1892) and George M. Minchin (1895) who performed the first "electrical" measurements of planetary and star light, respectively. The device they used was a photovoltaic cell that consisted of selenium (Se) deposited on an aluminum substrate. This device was employed by George F. Fitzgerald[6].

With a photoconductive Se cell, Joel Stebbins (1910) was able to achieve 2% precision on observations of Algol, and was able to discover the shallow secondary minimum in the narrow blue spectral region to which the detector was sensitive. Guthnik and Prager (1914) discovered the variability in β Cephei with the newly developed KH photoemissive cell.

Amplification of the minute currents produced by these cells was provided by Rosenberg (1929) and by Whitford (1932). The latter's thermionic valve DC amplifier became the basis for a rich spurt in observing. John Hall (1934) used a Cs-O-Ag cell to make the first near infrared measurements and found that cooling the detector with dry ice substantially improved the photometry. The great advantage of the RCA Cs-Sb photomultiplier was realized by Whitford and Kron (Whitford and Kron (1937)). For a full discussion of PEP electronic techniques and hardware, see the extensive article by Whitford (1962).

4.2 The Creation of Photometric Systems

The emergence of a new detector with all its advantages (linearity, high detective quantum efficiency, ...) does not automatically make a photometric system. We define a photometric system as

> *a calibrated subspace of magnitudes (or fluxes) and color indices (or flux gradients) where the zero points and scales of (each) magnitude and color have been carefully defined and calibrated by adequate (stellar) standards.*

Schwarzschild, in 1900, was one of the first to apply the concept of color index as we know it today: he coined the term *Farbentönung* as the difference between photographic and visual magnitude ($m_{\rm ph} - m_{\rm v}$). The concept led to a very straightforward application: *if* stars radiate as black bodies, *then* the flux distribution $\log I$ (as a function of λ^{-1}) is characterized by one single gradient, and

[6] Of the Lorentz–Fitzgerald contraction.

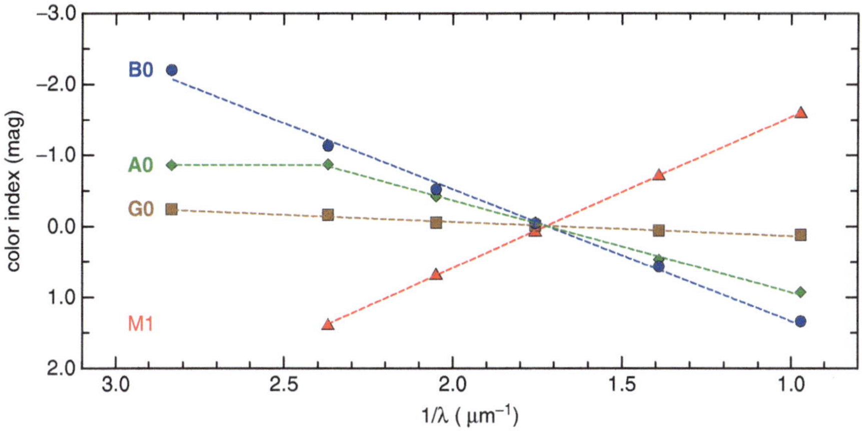

Fig. 5 Observed color index for main-sequence stars grouped by spectral type as a function of $1/\lambda$. The dashed lines are linear fits (for spectral type A0 the fit is constrained to $1/\lambda < 2.4$), and visualize black-body spectral energy distributions. Source: Sterken (2010)

hence the stellar spectrum is described by one single color index. In other words: *if the stellar energy distribution is commensurable with black-body radiation, only two passbands are necessary to describe adequately the stellar continuum.* Consequently, the greater the deviations from black-body radiation[7], the greater the number of passbands are needed and the higher the required order[8] of a photometric system.

Stebbins and Whitford (1945) photoelectrically determined the mean colors of unreddened main-sequence stars grouped by spectral type, and computed black-body colors for six wavelengths from Planck's formula. Figure 5 shows the measured color index for B0, A0, G0 and M1 main-sequence stars, and demonstrates that:

1. Color index (logarithm of the ratio of two stellar energy distributions) is almost linear in $1/\lambda$
2. Some stars show strong deviations, notably in the photometric band that contains the Balmer Jump
3. The color indices have a common zero point at $1/\lambda = 1.75$ (570 nm)

Astronomical photometry can thus also be perceived as *the determination, through the measurement of standardized magnitudes and color indices, of all deviations from black-body radiation*. However,

[7] Interstellar reddening-effects and spectral lines are also deviations from a black-body energy distribution.

[8] The order of a photometric system is the dimension of the vector space of significant and non-redundant magnitude and color-index parameters. With n bands, one gets $\frac{n(n-1)}{2}$ color indices.

- *to be of practical use*, such a system must be *transformable* – that is, it must remain internally consistent at any time, and provide homogeneous and uniform results wherever it is applied, and have the ability to reproduce magnitudes and color indices collected by an observer using equipment that is not the original equipment (or not identical to the original equipment), and
- *to be used effectively*, it should be supported by a database of measured objects that is considerably extensive, and made available and taught to potential users.

Naturally, the wisdom and erudition of the creators of the photometric system will, above all, be reflected in the merit of the end-product: *a technically superb photometric system will be of very limited use if it is based on an astrophysically inappropriate choice of passbands*. For example, knowing and understanding the origin of the Strömgren system could lead to less abuse of this powerful tool.[9]

Figure 6 shows the family tree of traditional photometric systems published by Sterken (1992a). The metaphor clearly illustrates that standard systems result from pruning and grafting the tree – in other words, that *maintenance of a standard system* is a prerequisite. This maintenance is not done by the users who consume the fruits, but by the gardeners, the *standardizers*.

The subsequent sections deal with four traditional photometric systems, viz., the Johnson, Strömgren, Geneva, and Walraven systems, through intercomparison of their passbands and their astrophysical applications for photometry and photometry-based spectral classification. Each of these sections has content that is also dealt with in the papers of Landolt, Wing, and Milone and Pel in this volume. In addition, infrared passband systems are discussed in the Milone and Young article, and many more optical-region passband systems are mentioned in Moro and Munari (2000).

4.2.1 The Johnson 3-Band System

Landolt, in this volume, deals with the creation of the Johnson *UBV* system; see also Wing's paper and Landolt (2007). The Johnson *B* and *V* filters were made from available glass to approximate the blue photographic response and the visual magnitude scale when combined with the 1P21 photomultiplier (a 9-stage, side-window type). It is very interesting to remember how this pioneer photometric system was conceived, as witnessed by Wisniewski's testimony (Wisniewski 1993):

> *"H. Johnson was well aware how the UBV filters should match the solar spectrum. But after the Second World War the number of available filters was limited. One would have had to pay for development and there were severe financial limitations. We could afford to spend just a few hundred dollars".*

---

[9] The system was designed to study mainly unreddened main sequence stars, but has been applied inaptly to objects all over the Hertzsprung–Russell diagram.

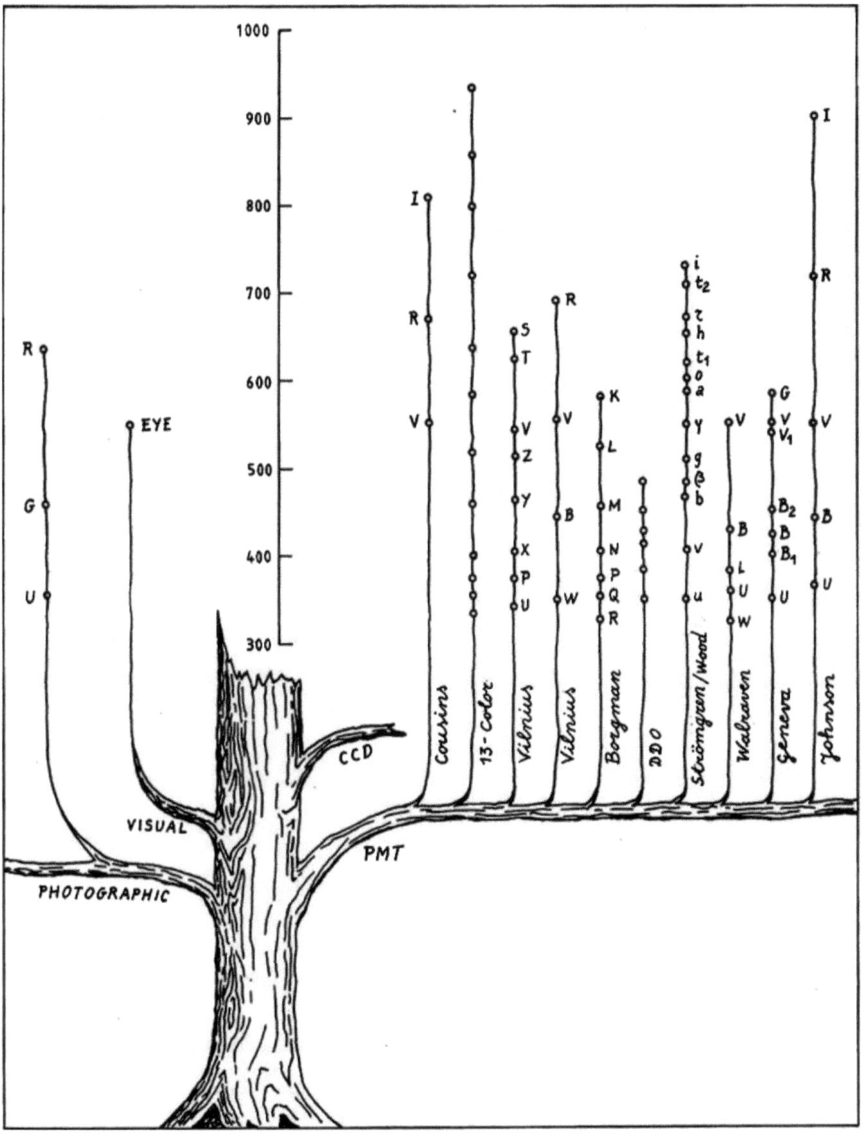

Fig. 6 Genealogy tree of classic photometric systems. Each main branch symbolizes a photometric detector, and the buds represent photometric bands arranged by wavelength along the vertical axis. The CCD branch sprout has developed to a full branch today. Reprinted from *Vistas in Astronomy*, Vol. 35, *On the future of existing photometric systems*, C. Sterken (1992a), with permission from Elsevier

Nevertheless, Johnson's original choices were mimicked in the design of quite a number of systems that were subsequently developed. A similar origin and result is cited for the Johnson *JHKLMNQ* passbands (see the Milone and Young paper in this volume and references therein).

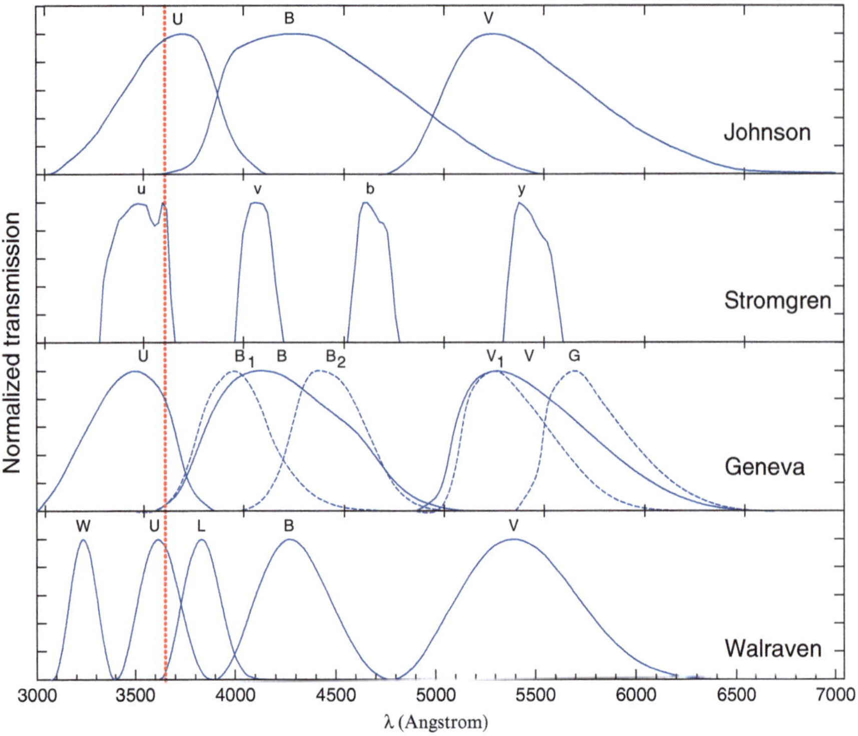

Fig. 7 Normalized bandpass response functions of *UBV*, *uvby*, $UBVB_1B_2V_1G$ and *VBLUW* systems. The dashed line indicates the position of the Balmer Jump

The normalized transmission curves of the Johnson, Strömgren, Geneva, and Walraven bands are shown in Fig. 7. The *UBV* system provides two color indices, viz., $B-V$ and $U-B$, which are very useful when plotting unreddened stars in an intrinsic two-color diagram (see Fig. 1 in Wing's paper for an example). The Johnson *UBV* system was extended to the red with the *R* and *I* bands (at 720 and 900 nm, respectively), hence the Johnson system is also referred to as the *UBVRI* system.

4.2.2 The Walraven 5-Band System

The most recent and quite complete description of the Walraven photometric system was given by Pel and Lub (2007); see also the Milone and Pel paper on page 45 in this volume. Figure 8 illustrates the tight setup of primary and secondary standards for Walraven photometry at La Silla. There are four aspects which make this system quite extraordinary, viz.,

1. It is the only system with an "ultraviolet" passband entirely on the short-wavelength side of the Balmer Jump (i.e., the *W* band, see Fig. 7)

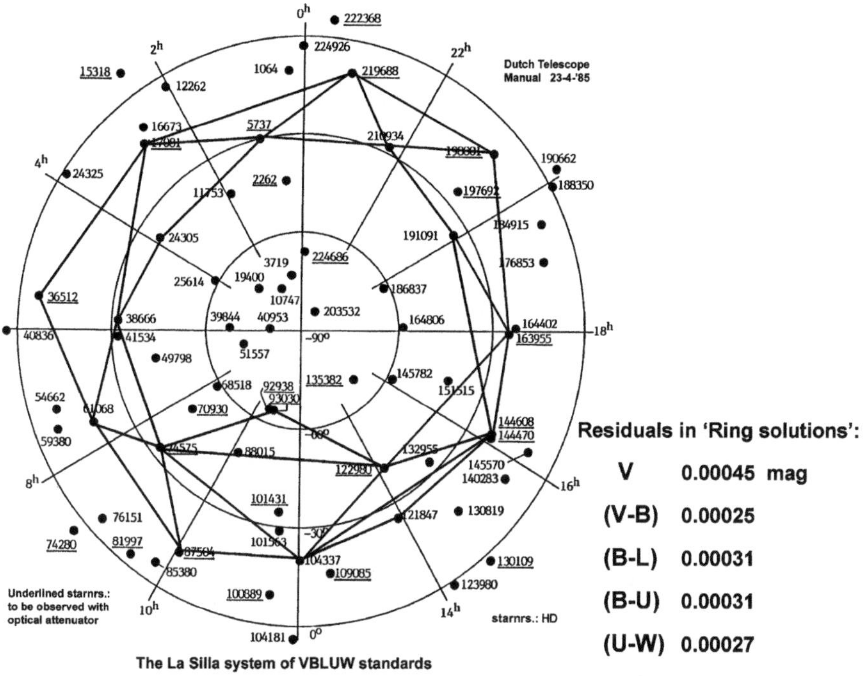

Fig. 8 The *VBLUW* standard stars used at ESO, La Silla. The heavy solid lines mark the pairs of 20 primary "ring standards" which form the basis of the *VBLUW* standard star calibration. The other 50 stars are the secondary standards which were tied to the primary ring (Pel and Lub 2007)

2. It is the only system which does not express light and color quantities in magnitude, but in the logarithm of the net detector signal, hence $m_{\mathrm{Walraven}} = \log d + c$ (where d stands for count rate, c for an arbitrary constant), unlike all other systems in which $m_{\mathrm{other}} = -2.5 \log d + c$
3. Only one Walraven photometer equipped with this passband system was ever developed; it was always used on the same telescope, first in South Africa, and later in Chile;
4. A supervised and centralized data reduction and standardization approach has always been taken; see Pel and Lub (2007) and also Fig. 8

Figure 9 shows Walraven's "light collector" equipped with his original photometer, and later with a CCD camera. The telescope was decommissioned in 1999, and recently transferred to *San Pedro de Atacama Celestial Explorations*[10] in northern Chile.

[10] SPACE, http://www.spaceobs.com/.

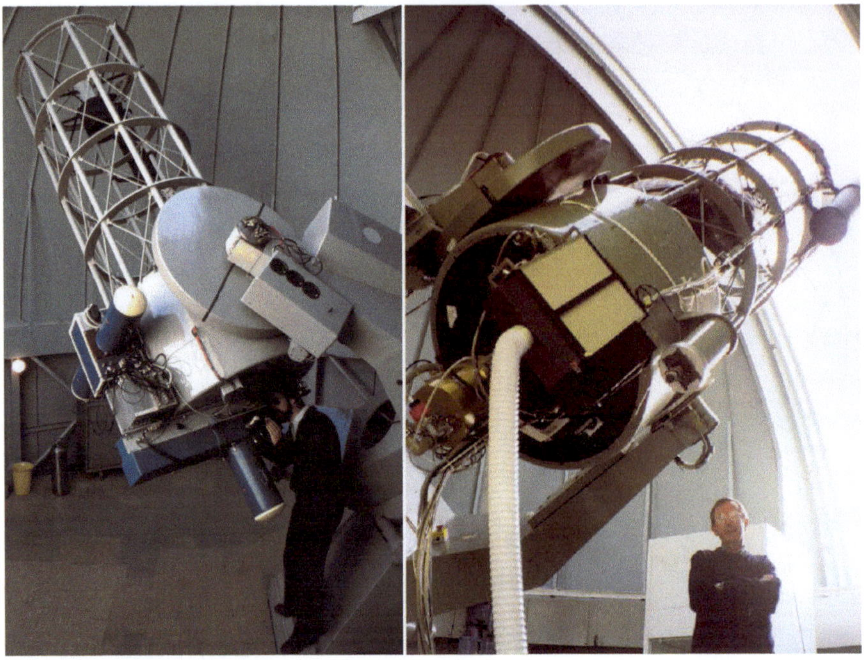

Fig. 9 The Dutch 91-cm telescope at ESO, La Silla. *Left*: with the Walraven five-channel photometer, shortly after their transfer to Chile in 1979. The 'L-shape' of the photometer optics layout can be recognized. The cylinder below the cassegrain focus is the coldbox for the L channel. The two other coldboxes for the (B,W) and (V,U) photomultipliers are mounted to the side of the telescope tube (source: Pel and Lub 2007). The observer is Jan Lub. *Right*: with CCD camera, together with the last observer (C. Sterken) on April 1, 1999, the day of decommissioning of the telescope (photo courtesy Lars Freyhammer)

4.2.3 The Strömgren *uvby*β System

The Strömgren system is an almost "filter-defined" system, using intermediate-width interference filters labeled y, b, v, u. The system, in fact, also incorporates two additional filters centered on the position of the Hβ line, making it a 6-band system, with five parameters: $c_1 = (u-v)-(v-b)$ (a measure of the Balmer Discontinuity), $m_1 = (v-b)-(b-y)$ (a measure of line blanketing, hence, metallicity), $b-y$ (an effective-temperature parameter for early-type stars), a magnitude y, and a β-index $\beta = -2.5\log\frac{N}{W} + c$, where N and W refer to, respectively, the narrow and wide bandpass of the Hβ filters. Note that y linearly correlates with V for most stars[11], but the Strömgren system does not possess its own standardized magnitude scale, hence *uvby* catalogs normally list only $b-y$, m_1 and c_1. The β index, because it involves filters at the same central wavelength, is not affected by atmospheric and

[11] Not for stars with peculiar spectra, such as luminous blue variables (LBVs), as noted on page 6.

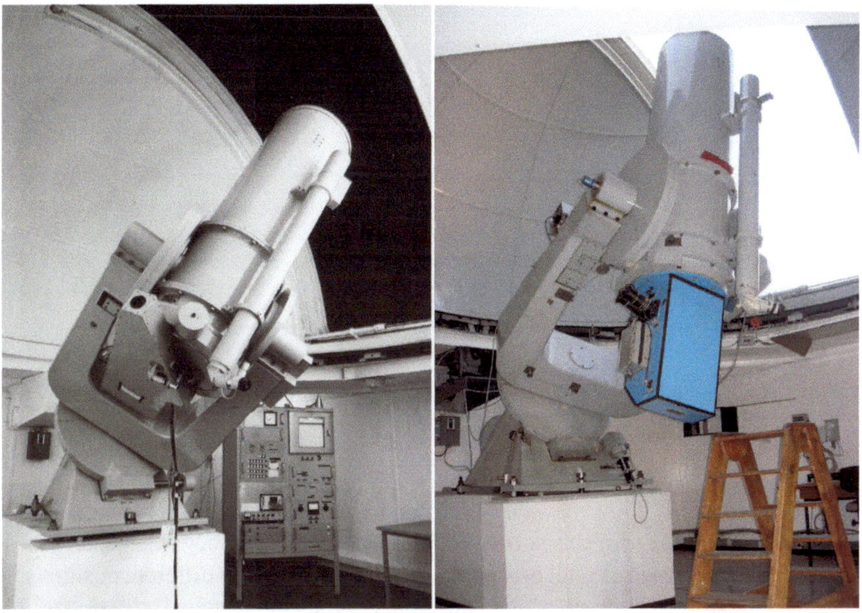

Fig. 10 *Left*: Danish 50-cm telescope and *uvby* photometer (1972, courtesy ESO). *Right*: Same telescope (SAT) with *uvby*β photometer (2010, photo C. Sterken)

interstellar extinction. The β index is very useful for deriving absolute magnitudes for early-type stars, but also for detecting emission features when compared with its Geneva-based counterpart, discussed in the next section.

There were two prototype designs of Strömgren photometers: the "flat" one described by Grønbech et al. (1976), and the more compact successor[12], mounted on the Danish 50-cm SAT (Strömgren Automatic Telescope); see Fig. 10. The photometer – in fact a spectrophotometer – is described by Florentin-Nielsen (1993). The wings of the *uvby* passbands, as shown in Fig. 7, are cut off by slots, as such rendering the passband edges very straight and very steep, a feature that may lead to problems in transformations.

Strömgren (1966) introduced two parameters that are unaffected by interstellar reddening[13]: $[c_1] = c_1 - 0.20(b-y)$ and $[m_1] = m_1 + 0.18(b-y)$. Figure 11 reproduces his original approach to *uvby*-based spectral classification, featuring the main sequence (spectral types B2 to K0) along the central band. For the sake of illustration, $[c_1], [m_1]$ data for F stars from the catalogue of Olsen (1994) have been overplotted. The comparison indicates that the application of these indices, especially for stars of later type, is very sensitive to the applied coefficient of $b-y$. Olsen (1979) uses 0.30 – and 0.32 in most of his later papers – a coefficient stemming

---

[12] A copy of which operates at San Pedro Mártir Observatory, Mexico.

[13] That obeys the standard reddening law.

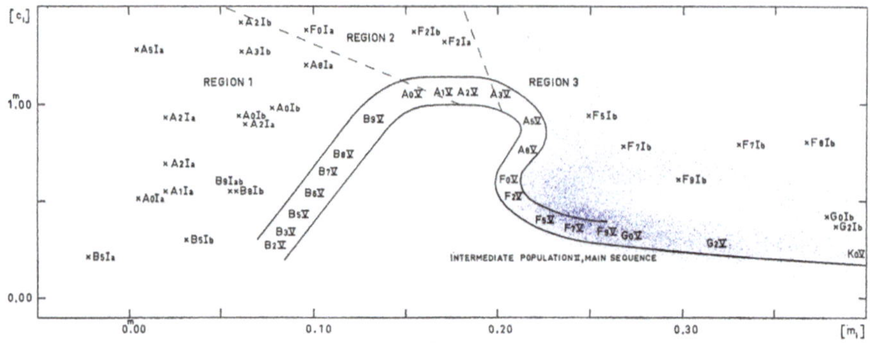

Fig. 11 $[m_1], [c_1]$ diagram from Strömgren (1966) with F stars from Olsen (1994) overplotted. The reddening-free m_1 index was calculated as $[m_1] = m_1 + 0.32(b-y)$ (see text)

from Crawford and Mandwewala (1976) and Crawford (1975). Here, again, is a clear example of the difference between precision and accuracy: the same data – thus of identical observational precision – lead to significantly different positions in the color-magnitude diagram when a different coefficient is applied.

It may seem surprising that the Strömgren system was not extended with one or two red bands, so to supply color indices that may be more suitable for late-type stars. The answer is in a comment by David L. Crawford on a paper by Michael S. Bessell (Crawford 1993b):

> *"Strömgren and I talked about red extensions to uvby, but decided that the existing R and I were perfectly adequate for any who needed a red extension, (e.g. for B, A and F stars). It was a conscious decision not to add two new filters; R, I are, in fact, very useful supplements to uvby for many applications".*

4.2.4 The Geneva 7-Band System

The Geneva seven-color photometric system, developed by Marcel Golay in the late 1950s, was intended to duplicate the Johnson *UBV* system with a similar *V* band, but slightly different *U* and *B*. To these three basic bands were added four additional intermediate-bandwith filters (B_1, B_2, V_1, G) with properties comparable to the Barbier–Chalonge–Divan system of stellar classification, the so-called *BCD* spectral classification system (Chalonge and Divan 1973). The description of the original Geneva system and the methodology of data reduction was published in Rufener's 1964 PhD thesis "*Technique et Réduction des Mesures dans un Nouveau Système de Photométrie Stellaire*",[14] which clearly describes the hardware, the measurements, the data reduction and the list of standards (Rufener 1964). Rufener also devised a modified Bouguer method – the M and D technique – for extinction

[14] Technique and Reduction of Measurements in a New (Stellar) Photometric System.

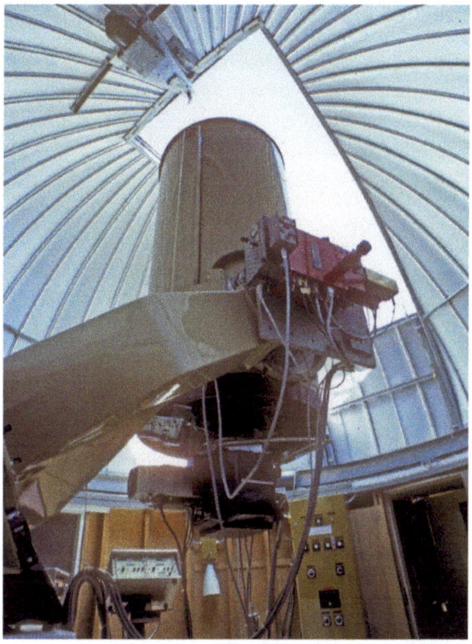

Fig. 12 Swiss 70-cm telescope at ESO La Silla. Photo courtesy N. Cramer

determination using simultaneously ascending and descending stars. The Geneva color indices lead to a set of three orthogonal reddening-free parameters X,Y,Z – for a definition we refer to Cramer and Maeder (1979) and Cramer (1993). An interesting aspect of Geneva photometry is the correlation with the Strömgren reddening-free β index through a polynomial estimator (Cramer 1984): $\beta(X,Y)$, where X,Y measures the spectral continuum, and β measures the strength of the Hβ line. The difference $\beta(X,Y) - \beta$ is very sensitive to emission in Hβ in O, B and early-type A stars. Figure 12 shows the Swiss 70-cm telescope at ESO La Silla.

Geneva and Johnson photometry is mutually transformable at about the 1% level (at best). The Geneva system is characterized by several outstanding aspects, viz.:

1. High homogeneity
2. Transmission functions of the filters (all glass) are almost invariant with time
3. Temperature-controlled detectors
4. It is truly a "filter-defined" system (as opposed to, say, *UBV*)
5. Stringent procedure for reductions to outside the atmosphere
6. Tightly supervised data reduction procedures

4.3 *Open and Closed Photometric Systems*

The Walraven and Geneva systems are called "closed" systems, i.e., they have been deployed in only a small number of photometers (even only 1), they are used by a small group of observers, and are subject to a unique data reduction methodology.

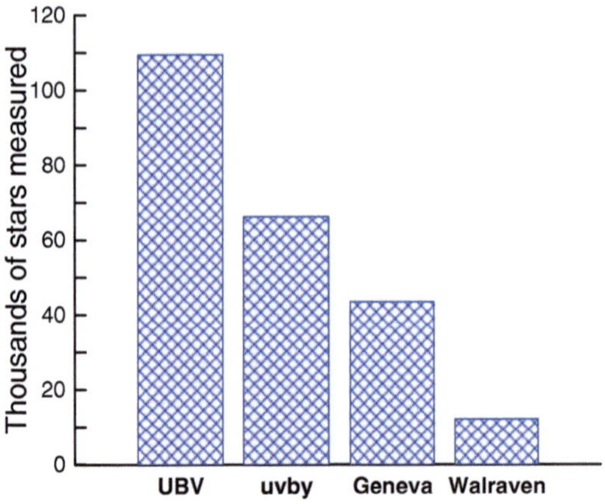

Fig. 13 Approximate number of data in *UBV*, *uvby*, $UBVB_1B_2V_1G$ and *VBLUW* catalogs. Based on data from the *General Catalogue of Photometric Data*, http://obswww.unige.ch/gcpd/

The two others, *UBV* and *uvby*, on the contrary, are "open" systems, reflected in the number of measured stars, see the bar chart in Fig. 13. Rufener (1985) distinguishes between "open" and "closed" calibrations, which widens the concepts to include all aspects of data reduction, standardization and calibration, He tabulates the options which distinguish both orientations, viz.:

Natural system	Open	Closed
Great variety	Stable	Supervised
Data acquisition	Heterogeneous	Complete homogeneity
Data reduction	Heterogeneous	Complete homogeneity
Data releases	Dispersed in literature	Single source of compilation
Calibration parameters	of general consensus	from a small number of primary standards
Calibration form	Generic correlations	validated calibration
Persons involved	$\geq$100	maybe $\leq$10
Discrimination performance	rough	much finer

Evidently, the "closed" approach yields smaller databases, but its tighter stability is not just the result of having only one telescope and one photometer. Figure 14 illustrates this clearly: the Geneva system has been developed over more than half a century, with seven different photometers – although with filters cut from the same original glass sheets – at half a dozen observing sites (at elevations ranging from 650 to 3,500 m) and different telescopes. Nevertheless, the system has kept tight standards of operation. The timeline clearly illustrates the progressive deployment of each new photometer as it became available. Note that this timeline covers only classical Geneva photomultiplier tube (PMT) photometry, and excludes the Geneva CCD branch.

Figure 15 shows a selection of early B-type stars (giants and subgiants) that appear in two Geneva catalogs (Rufener 1980, 1988), see the discussion in Sterken (1993a). The full lines are the calibrations for spectral clases V, III and Ia from Cramer (2005).

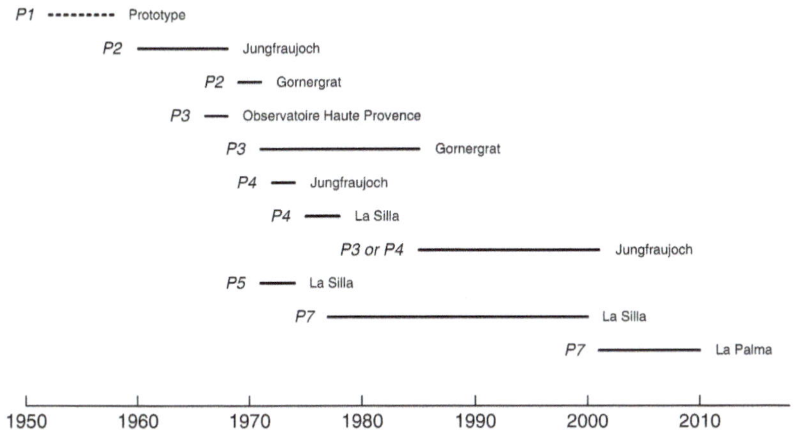

Fig. 14 Timeline of Geneva photometers P1–5 and P7, and their locations. P6 was never deployed on an observing site (based on Cramer (2005, 2010) Personal Communication)

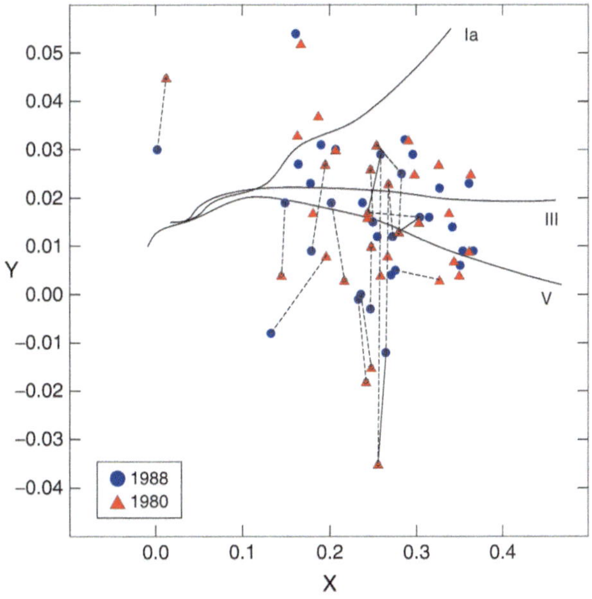

Fig. 15 Position of all β Cephei stars with Geneva photometric indices listed in the catalogues of Rufener (1980) and Rufener (1988). Data couples for which the differences exceed the quoted accuracy of Geneva all-sky photometry are connected with dashed lines. The full lines are the calibrations for spectral classes V, III and Ia. After Sterken (1993a)

5 Improved Photometry Through the CCD Era

This section is kept short because a good historical review is given in Howell's paper in this volume, and in Howell's book (Howell 2006), which is also a good source for in-depth discussion of methods.

It may be argued that it is of little use to concentrate on guidelines to improve on PMT-based photometry: very few such classical photometers remain in function, and the younger generation of users is less familiar with their operation. CCDs have obvious advantages (although admittedly many disadvantages): many objects can be recorded in one exposure, with a deeper limiting magnitude, and with much less concern for sky background and transparency variations (some of the disadvantages are noted below). We henceforth concentrate on discussing the limitations of photometry with an eye to improving the precision and accuracy of CCD photometry.

5.1 Removal of Systematic Effects

Some sources of systematic error that bedevil photoelectric photometry remain – and may get even more serious – in CCD applications. To mention only one, passbands are now a combination of another detector response coupled to passband characteristics of imaging filters.[15] Consequently, it is worthwhile to mention the precautions that should be taken before any modern precise (and hopefully accurate) multi-band photometry can be undertaken (see also Sterken 1995). In particular, the photometrist should ensure or be aware that:

1. All filters have the same imaging quality characteristics
2. The filter wheel is mounted perpendicular to the optical axis of the telescope – especially when using interference filters, as their optical characteristics are more strongly dependent on the angle of incidence than is the case with glass filters, see Young (1974)
3. Observers should have a proper level of training and experience
4. The signal-to-noise ratio is not constant over a CCD frame because the integration time of all objects is the same – but not their brightnesses
5. Some CCDs have high reflectivity, with the reflected light getting back to the detector, often in an unwanted spectral region (red and blue leaks), which leads to ghosts, and thus contributes to local photometric errors
6. Cosmic-ray events are properly dealt with
7. Bias and dark data are adequate, and subtraction is performed optimally
8. Any PSF variation is treated adequately

[15] One must not forget that a PMT-based photometer is a 1-pixel imager: the Fabry lens concentrates all light in one spot of homogeneous light, whereas the CCD camera can be a multimillion pixel imager.

9. The non-uniform response inside a pixel (by as much as 10%) is understood and dealt with
10. Twilight flatfielding, except at very high latitude sites, where the twilight interval is extended, poses two problems due to fast-changing twilight conditions:

    a. Short exposures are inevitable, and the flatfield structure will be significantly altered by the shutter function
    b. Only very little time is available for taking the frames, so that the nightly number of frames is likely to be small

11. Shutter timing effects, viz., the (spacially non-uniform) differences among actual and presumed exposure times of objects are understood (this problem alone can ruin any calibration at the millimagnitude level)
12. Undersampling, when it occurs, is understood: the small size of some chips forces the use of a scale that allows large sky coverage and leads to undersampling. In order to cover star images with several pixels and to fully utilize any good seeing, a pixel size no greater than $0''.5$ is desired. With pixel sizes ranging from 20 to 30 μm that implies a focal length of 10–15 m. Photometric accuracy falls off rapidly as the star images become undersampled;
13. CCDs are now approaching the size of small photographic plates, and because the FWHM can be 10% larger at the corner of a 13 arcmin square field (CTIO 0.9-m $f/13.5$ telescope, Walker 1993), a field flattener may be needed to avoid a strongly varying PSF as a function of field position. The same problem can occur if the CCD chip itself is not flat (as with early photographic plates);
14. Detector temperature is known and controlled. Most CCDs are linear to at least 0.1–0.5% from very low signals to at least 80% of the full well capacity, yet a surprising number of CCD instruments show non-linearity and other problems of the sort long familiar to photographic photometrists (Young 1994). However, linearity is usually assumed, based on the absence of apparent adverse effects, but vigorous linearity tests seem to be carried out rarely. It would be highly valuable if all observing sites provided a robust linearity-check routine at the telescope;
15. Filter temperatures are known and controlled. Glass-filter temperature coefficients are much larger than detector temperature coefficients, and even interference-filter temperature coefficients are comparable to those of CCDs, see Young (1967).

5.2 Extinction Corrections

Correction for atmospheric extinction is the first step in the data reduction process. Extinction correction, in the past, was always taken seriously in the sense that considerable time was spent on the measurement of extinction stars during each observing night; that does not necessarily mean that optimal procedures for extinction determination were always followed. Photomultiplier photometrists would usually measure one or several "extinction stars" (bright, non-variable stars of different colors) at a range of airmasses (measurements near the meridian, and at airmasses 1.3, 1.6, 1.9, 2.0) and then derive the nightly extinction through a linear regression of apparent magnitude on the airmass X. Very little attention was given to variable extinction, and to east–west asymmetries at the observing site. Explicit observations for the determination of extinction are nowadays done less and less since the CCD detector became the dominant photometric detector. The reason not only is the prevailing (but wrong) opinion that differential CCD magnitudes need no extinction correction, but also the fact that fewer and fewer observers spend sparse observing time on extinction and standardization measurements.

There are three aspects to the extinction problem: the observation of extinction star(s), the extraction of the extinction coefficient(s), and the temporal behavior of the extinction phenomenon itself.

5.2.1 Observation of Extinction Stars

Young (1974) presents several examples of Bouguer plots showing east-west differences due to a horizontal extinction gradient, due to time-varying extinction, and also because of instrumental drifts. The main concern was that extinction star measurements would be performed several times each night.

Rufener (1964), see also Rufener (1985), developed the extinction-star observation strategy to a higher level by introducing his concept of M and D stars. Remarking that the Bouguer approach rests on the simultaneous satisfaction of three hypotheses, viz.:

1. Constancy of the extinction star
2. Stability of the photometer response
3. Constancy and isotropy of the atmosphere during many hours of observation

he damage-controls each hypothesis by restriction: selecting stable stars, advanced stabilization of the instrument, and relaxing the third element by assuming only isotropy of the atmosphere. The extinction variations are then eliminated by observing quasi-simultaneously an ascending extinction star M and a descending[16] extinction star D. This so-called $M+D$ method has been applied for decades in Geneva photometry.

A procedure similar to Rufener's $M+D$ method is available in the `PEPSYS`[17] contribution to ESO-Midas, through the planning program to select extinction-star timings, and by using the reduction program – which allows fitting at least a linear trend to the extinction with time.

---

[16] The symbols come from the first letter of the French words for ascending and descending: *Montante, Descendante.*

[17] ESO-MIDAS User's Guide Volume B, the *"PEPSYS general photometry package"*, http://www.eso.org/sci/data-processing/software/esomidas//doc/user/98NOV/volb/node248.html.

5.2.2 Extraction of Extinction Coefficients

A crucial variable in the determination of the extinction coefficient is the airmass[18] X. Young (1974) pointed out that mis-using Bemporad's (1904) table (by assuming that the argument is true rather than refracted zenith distance) produces an error of 0.01 airmass at $sec z = 2.65$ which can lead to millimagnitude inaccuracies when combining measurements taken in the meridian and at high airmasses. Young thus proposes to use the formula $X = \sec z[1 - 0.0012(\sec^2 z - 1)]$ which is accurate up to $sec z = 4$. This formula, unfortunately, was copied by Hall and Genet (1982) with the omission of the exponent 2 in the argument sec z, which has led to programming of this error in photometric reduction codes (Tuvikene (2008) Personal communication).

The idea of determining the extinction coefficient from measurements of many stars – instead from just one or two – was already discussed by Gheury (1913) when applied to a set of 100 circumpolar stars at Harvard College. The multi-star method (for more details, see Sterken and Manfroid (1992b), is widely applied in all-sky photometry, and also works very well in CCD photometry of star clusters. Schwarzenberg-Czerny (1991) also outlines a similar method that leads to improved statistical errors.

As noted above, differential CCD photometry is carried out often without any care for extinction determination. This is regrettable, because time-series monitoring can provide the user with the free bonus of accurate extinction information, especially for stars in clusters. Even the simple Bouguer plot reveals at one glance whether the night was very good (tightly linear magnitude versus airmass relation) or poor (bifurcated Bouguer line, jumps or loose scatterplot). It is a matter of considerable regret that this simple verification more often than not is forgotten.

5.2.3 Atmospheric Extinction in Time

The recording – more important even the publication *in extenso* – of extinction coefficients serves two purposes, viz.:

1. Qualitative comparison of observing sites (i.e., an element of site testing, see Fig. 16)
2. Monitoring of the temporal stability of atmospheric extinction at a given site (again with a useful application to site evaluation)

Gutierrez-Moreno et al. (1982) analyzed the extinction behavior at Cerro Tololo covering the time interval 1964–1980. Figure 17, from Sterken et al. (1986), illustrates the difference between the extinction behavior at two excellent sites: La Silla Observatory (Chile) and Mount Laguna Observatory (California, U.S.) seven years later. Note that the difference is not per se an indicator of site quality, but rather reflects a worldwide phenomenon attributed to the presence of long-living

[18] Note that here we represent airmass by X, although M has been used also elsewhere.

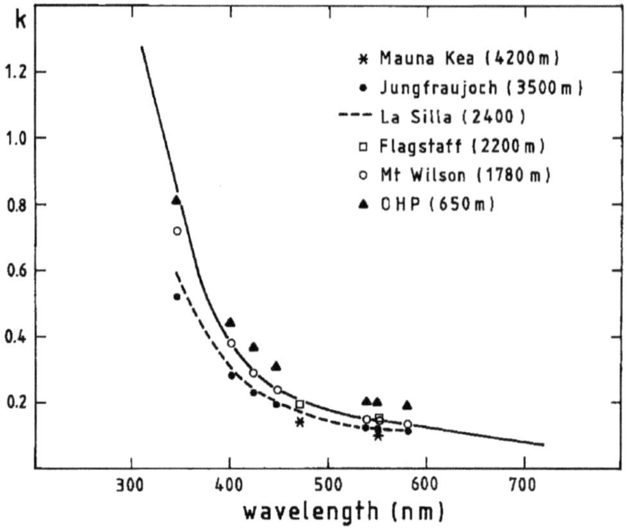

Fig. 16 Atmospheric extinction coefficient for selected sites. The full line is based on data from Melbourne (1960). Source: Sterken and Manfroid (1992b)

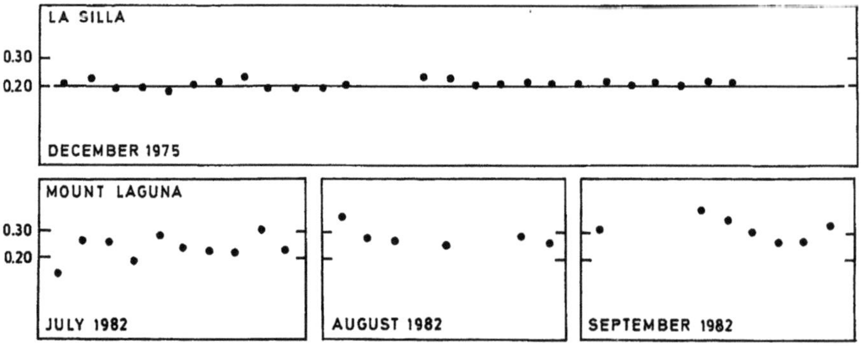

Fig. 17 The Strömgren *b* monochromatic extinction coefficient for La Silla (1975) and Mount Laguna (1985) observatories. Source: Sterken et al. (1986)

dense stratospheric aerosols produced by the eruption of the El Chichón volcano on March 29 and 3–4 April 1982. These authors point out the dangers of not using extinction coefficients determined on a nightly basis. Poretti and Zerbi (1993a,1993b) describe spurious effects produced by variable extinction coefficient in photoelectric photometry.

Schuster et al. (2002) present extinction data covering two decades of observations in the 13-color system and in the *uvby* system, together with a description of their observing procedures. Rufener (1986) also describes a decade of photometry at La Silla, and concludes that the period 1978–1982 had the best atmospheric transparency within two decades. Unfortunately the onset of the El Chichón event was

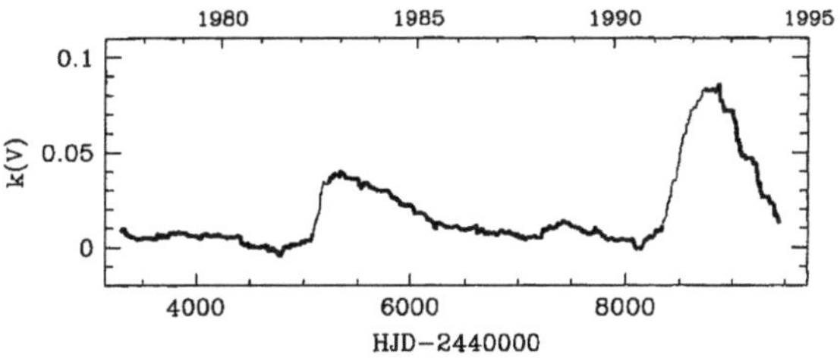

Fig. 18 Variation of the part of the extinction (in Geneva V) due to the volcanoes El Chichón and Mount Pinatubo. Source: Burki et al. (1995)

missed because the observers worked in only "constant airmass mode." Sterken and Manfroid (1992a) described *uvby* extinction data covering more than 15 years of observations at ESO La Silla, revealing three periods of increased extinction due to airborne aerosols of volcanic origin. Figure 17 illustrates the effect in the Strömgren *b* band. Rufener's (1986) paper was followed by a sequel by Burki et al. (1995) covering about 4,400 nights (1975–1994). This paper describes the timescales of cyclic variations, and also of the episodes of volcanic perturbations. Figure 18 shows the magnitude of the impact of the eruption, the time delay for the volcanic aerosols to reach the observing site, and the duration of the perturbations.

5.3 *Transformations*

This subject has been discussed in detail by, for example, Sterken (1993b), so here we provide only a brief overview of what is involved, and describe ways in which greater precision can be achieved.

The importance of derivatives of the stellar spectral flux was considered by Strömgren (1937) and by King (1952) and further developed by Young (1992). The latter made use of concepts of functional analysis (see Putnam 1967, Oden 1979, or Kolmogorov and Fomin 1999). Basically the idea, explained clearly in Young (1993), is to consider spectral irradiances as vectors in Hilbert space. These vectors undergo rotations as they are transformed by the atmosphere and further by the atmospheric transmission functions, the filter passband, and the sensitivity curves of the telescope optics and the detector. These ideas have been applied to establish the IRWG passband system discussed by Milone and Young in this volume.

The main result is that two systems are exactly transformable if, and only if, the passbands of one are a linear combination of the passbands of the other. In general, this condition cannot be met; but transformation errors are minimized by the linear combinations of passbands that minimize the fitting errors in the least-squares sense. Essentially, the problem is to synthesize the response functions of the target system numerically.

If the measured intensities are linear, it makes no difference whether the linear transformations are applied to the passbands or to the observed intensities; transforming intensities is equivalent to transforming the spectral response functions. The classical example is that given by Johnson (1952), who showed that the broad passbands of the old International Photographic system could be adequately approximated by a linear combination of the intensities measured through the newly introduced U and B filters.

Notice that strictly accurate transformation can be done only in terms of intensities, not magnitudes. The empirical "transformations" traditionally used, which involve magnitudes and color indices, are interpolation formulae based on the unrealistic assumption that stellar spectra are like those of black bodies (see Fig. 5). In fact, as Young (1992) showed, such interpolation formulae rely on series expansions that diverge.

If stars were really black bodies, multicolor photometry would be redundant. In reality, stars are not black bodies, and real deviations from this ideal allow the separation of interstellar reddening from "temperature reddening", besides other useful information. But this information becomes garbled by the nonlinear logarithmic transformation of intensities into magnitudes; then transformations involving magnitudes and color indices become nonlinear and multi-valued, and systematic errors that depend on reddening, metallicity, and other astrophysically important effects contaminate the results.

Obviously, passband transformations are most accurate if the individual response functions are smooth and overlapping. For best results, the peak of one band should fall on the steepest slopes of its neighbors. The steep edges produced by sharp-cutoff glass filters and multi-cavity interference filters should be avoided. Unfortunately, these requirements for accurate transformability are not met by existing photometric systems.

5.4 *Prescriptions and Guidelines for Photometric Accuracy*

Photometric accuracy (at a good site) depends on three factors: the quality of the detector and its control system, the observing and data-reduction procedures, and the design of the passbands. The first two factors are in constant progress, the latter point really is the bottleneck of photometry.

The basic prescription for photometric passbands is to keep each response vector as close as possible to a projection on a standard system. This is accomplished in the following ways:

- Treat extinction as a transformation
- Allow passbands of one system to be represented as linear combinations of passbands to create transformability between systems

These steps require redesign of passbands (see the Milone and Young paper in this volume for an example of the successful implementation of this idea in the infrared). The redesign of passbands in turn, requires the following prescription for success:

- Apply gentle slopes to the spectral edges of the filters, and avoid "rippled tops"
- Space passbands so that peaks fall at the steepest parts of neighboring passbands
- In the visual (or optical) region, passbands should be overlapping

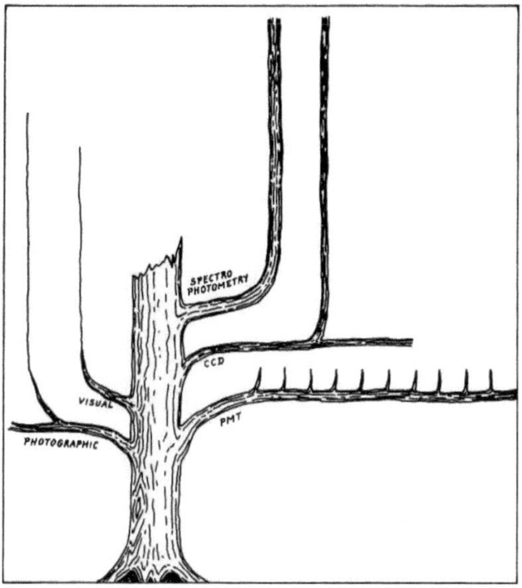

Fig. 19 A dream for the future: a few existing photometric systems will survive, but most of them will disappear. The CCD-branch should yield only one single-designed photometric system, alongside synthetic photometry on the basis of spectrophotometric data. Reprinted from *Vistas in Astronomy*, Vol. 35, *On the future of existing photometric systems*, C. Sterken (1992b), with permission from Elsevier

In designing new systems and instruments, one should try to consult colleagues who have developed succesful setups in the past, the way Crawford (1993a) stated:

> *"I have also had the pleasure of talking photometry with many of the pioneers (and the new photometrists too), such as Stebbins, Whitford, Kron, Johnson, Harris, Stromgren, Cousins, and with others with great insight – such as Morgan and Garrison relative to classification matters ... "*

without making the error of slavishly copying poorly conceived passbands just because a large body of data has been taken in them.

Figure 19 is reproduced from Sterken (1992a) and is the dream vision of how photometric systems should develop. The reality, almost two decades later, is very different, and unfortunately seems to develop in exactly the same way as photoelectric systems developed since the 1950s.

Acknowledgements It is a pleasure to thank Dr. N. Cramer for providing useful information that allowed us to construct the timeline given in Fig. 14.

References

Baum, W. A. (1962). In W. A. Hiltner (ed.), *Astronomical Techniques*, in *Stars and Stellar Systems* series, Vol. **II**, (Chicago: University of Chicago Press), pp. 1–33

Bemporad, A. (1904). Heidelberg Mitt. **4**

Bond, G. P. (1859). *Astronomische Nachrichten*, **49**, 81

Burki, G., Rufener, F., Burnet, M., et al. (1995). *Astronomy and Astrophysics, Supplements*, **112**, 383

Butler, C. J., & Elliott, I. (1993). *Stellar Photometry*, (Cambridge: Cambridge University Press), p. 3
Cameron, D. M. (1951). *Astronomical Journal*, **56**, 92–96
Chalonge, D., & Divan, L. (1973). *Astronomy and Astrophysics*, **23**, 69
Cramer, N. (1984). *Astronomy and Astrophysics*, **141**, 215
Cramer, N. (1993). *Astronomy and Astrophysics*, **269**, 457
Cramer, N. (2005). *Orion*, **326**, 8
Cramer, N., & Maeder, A. (1979). *Astronomy and Astrophysics*, **78**, 305
Crawford, D. L. (1975). *Publications of the Astronomical Society of the Pacific*, **87**, 481
Crawford, D. L. (1993a). In D. Kilkenny, E. Lastovica & J. W. Menzies (eds.), *Precision Photometry*, (Capetown: South African Astronomical Observatory), p. 18
Crawford, D. L. (1993b). In C. J. Butler & I. Elliott (eds.), *Stellar Photometry*, (Cambridge: Cambridge University Press), p. 38
Crawford, D. L., & Mandwewala, N. (1976). *Publications of the Astronomical Society of the Pacific*, **88**, 917
Coblentz, W. W. (1912). *Bulletin of the Bureau of Standards*, **9**, 15
Davis, A., Barkume, K., Springob, C., et al. (2004). *Journal of the American Association of Variable Star Observers*, **32**, 117
Eichner, J., et al. (1947). *Astronomical Journal*, **53**, 25
Florentin-Nielsen, R. (1993). In C. J. Butler & I. Elliott (eds.), *Stellar Photometry – Current Techniques and Future Developments*, (Cambridge: Cambridge University Press), p. 213
Garrison, R. F. (1993). In C. J. Butler & I. Elliott (eds.), *Stellar Photometry*, (Cambridge: Cambridge University Press), p. 39
Grønbech, B., Olsen, E. H., & Strömgren, B. (1976). *Astronomy and Astrophysics, Supplement*, **26**, 155
Gheury, M. E.-J. (1913). *Ciel et Terre*, **34**, 351
Gutierrez-Moreno, A., Cortes, G., & Moreno, H. (1982). *Publications of the Astronomical Society of the Pacific*, **94**, 722
Guthnik, P., & Prager, R. (1914). *Veröffentlichungen der königlichen Sternwarte Berlin–Babelsberg*, **1**, 1
Hall, J. S. (1934). *Astrophysical Journal*, **79**, 145
Hall, D. S., & Genet, R. M. (1982). *Photoelectric Photometry of Variable Stars*, (Fairborn, Ohio: Fairborn Press), p. 9–4
Hearnshaw, J. B. (1996). *The Measurement of Starlight: Two Centuries of Astronomical Photometry*, (Cambridge: Cambridge University Press)
Hearnshaw, J. B. (2000). "Nineteenth Century Visual Photometers and Their Achievements," in C. Sterken & K. Staubermann (eds.), *Karl Friedrich Zöllner and the Historical Dimension of Astronomical Photometry*, (VUBPress), p. 19
Howell, S. B. (2006). *Handbook of CCD Astronomy* (2nd ed.), (Cambridge: Cambridge University Press)
Hu, J. Y., de Winter, D., Thé, P. S., & Perez, M. R. (1990). *Astronomy and Astrophysics*, **227**, L17
Johnson, H. L. (1952). *Astrophysical Journal*, **116**, 272
Jordan, F. C. (1923). *Astronomical Journal*, **35**, 44
Jordan, F. C. (1929). Publs., Allegheny Observatory, **7**(2), 125–130
Kibblewhite, E. J. (1971). In H. Seddon & M. J. Smyth (eds.), *Automation in Optical Astrophysics*. Proc. IAU Colloq. 11, *Publications of the Royal Society of Edinburgh*, **8**, p. 122
King, I. (1952). "Effective Extinction Values in Wide-Band Photometry," *Astronomical Journal*, **57**, 53
Kolmogorov, A. N., & Fomin, S. V. (1999). *Elements of the Theory of Functions and Functional Analysis*, (Mineola, NY: Dover Publications)
Landolt, A. U. (2007). In C. Sterken (ed.), *The Future of Photometric, Spectrophotometric and Polarimetric Standardization*, ASP Conference Series, Vol. **364**, p. 27
Majewski, S. R. (1992). *Astrophysical Journal*, **78**, 87
Melbourne, W. G. (1960). *Astrophysical Journal*, **132**, 101

Meyermann, B., & Schwarzschild, K. (1905). *Astronomische Nachrichten*, **170**, 277–282
Milone, E. F. & Pel, J. W. (2011). "The High Road to Astronomical Photometric Precision: Differential Photometry," in E. F. Milone (ed.), *Astronomical Photometry*, (New York: Springer), p. 33
Minchin, G. M. (1895). *Proceedings of the Royal Society*, **58**, 142
Monck, W. H. S. (1892). *Astronomy and Astrophysics*, **11**, 843
Müller, G., & Kempf, P. (1899). *Astronomische Nachrichten*, **159**, 203
Moro, D., & Munari, U. (2000). *Astronomy and Astrophysics, Supplement*, **147**, 361
Oden, J. T. (1979). *Applied Functional Analysis*, (Englewood Cliffs, NJ: Prentice-Hall)
Olsen, E. H. (1979). *Astronomy and Astrophysics, Supplement*, **37**, 367
Olsen, E. H. (1994). *Astronomy and Astrophysics Supplement Series*, **106**, 257
Parkhurst, & Jordan (1907). *Astrophysical Journal*, **26**, 244
Pel, J. W., & Lub, J. (2007). In C. Sterken (ed.), *The Future of Photometric, Spectrophotometric and Polarimetric Standardization*. ASP Conference Series, Vol. **364**, p. 63
Pickering, E. C. (1888). *Memoirs of the American Academy of Sciences*, **11**, 179
Plaut, L. (1940). *Bulletin of the Astronomical Institutes of the Netherlands*, **9**, 121
Pogson, N. (1856). *Monthly Notices of the Royal Astronomical Society*, **17**, 12
Poretti, E., & Zerbi, F. (1993a). *Astronomy and Astrophysics*, **268**, 369
Poretti, E., & Zerbi, F. (1993b). "Instantaneous Determination of a Variable Extinction Coefficient," in C. J. Butler & I. Elliott (eds.), *Stellar Photometry* (p. 221), (Cambridge: Cambridge University Press), p. 31
Pritchard, C. (1885). *Uranometria Nova Oxoniensis: Astronomical Observations made at the Oxford Observatory*, No. II
Putnam, C. R. (1967). *Commutation Properties of Hilbert Space Operators, and Related Topics*, (New York: Springer-Verlag)
Rosenberg (1929). *Handbuch der Astrophysik*, **2**, Part 1, 380
Rufener, F. (1964). *Technique et Réduction des Mesures dans un Nouveau Système de Photométrie Stellaire*, Observatoire de Genève
Rufener, F. (1980). *Third Catalogue of Stars Measured in the Geneva Observatory Photometric System*, (Genève: Observatoire de Genève)
Rufener, F. (1985). In *Calibration of Fundamental Stellar Quantities*, (Dordrecht: D. Reidel Publishing Co.), pp. 253–268
Rufener, F. (1986). *Astronomy and Astrophysics*, **165**, 275
Rufener, F. (1988). *Catalogue of Stars Measured in the Geneva Observatory Photometric System*, Observatoire de Genève (4th ed.)
Schaefer, B. (2010). *Astrophysical Journal, Supplement*, **187**, 275–373
Schilt, J. (1922). *Bulletin of the Astronomical Institutes of the Netherlands*, **1**, 51
Schuster, W. J., Parrao, L. & Guichard, J. (2002). *The Journal of Astronomical Data*, **8**, 2
Schwarzenberg-Czerny, A. (1991). *Astronomy and Astrophysics*, **252**, 425
Siedentopf, H. F. (1935), *Astronomische Nachrichten*, **254**, 33–39
Seares, F. H., & Joyner, M. C. (1943). *Astrophysical Journal*, **98**, 261
Seares, F. H., & Joyner, M. C. (1945). *Astrophysical Journal*, **101**, 15
Seares, F. H., Ross, F. E. & Joyner, M. C. (1941). *Magnitudes and Colors of Stars North of* $+80°$, (Washington, D.C.: Carnegie Institution Publication), No. 532
Stebbins, J. (1910). *Astrophysical Journal*, **32**, 185
Stebbins, J., & Whitford, A. J. (1945). *Astrophysical Journal*, **102**, 318
Sterken, C. (1992a). "On the Future of Existing Photometric Systems," *Vistas in Astronomy*, **35**, 139
Sterken, C. (1992b). "On the Impact of an Ultimate Metrological Reform in Astronomical Photometry," *Vistas in Astronomy*, **35**, 263
Sterken, C. (1993a). "Irreducible Elements of Quality in High-Precision Photometry," in C. J. Butler & I. Elliott (eds.), *Stellar Photometry*, (Cambridge: Cambridge University Press), p. 92

Sterken, C. (1993b). "On Photometric Precision, Error and Uncertainty on Time Scales of Decades," in D. Kilkenny, E. Lastovica & J. W. Menzies (eds.), *Precision Photometry*, (Capetown: South African Astronomical Observatory), p. 57

Sterken et al. (1999). *Astronomy and Astrophysics*, **346**, L33

Sterken, C. (2010). "Writing a Scientific Paper II. Communication by Graphics," in C. Sterken (ed.), *Scientific Writing for Young Astronomers* EDP Sciences, p. 222

Sterken, C., & Manfroid, J. (1992a). *Astronomy and Astrophysics*, **266**, 619

Sterken, C., & Manfroid, J. (1992b). *Astronomical Photometry, A Guide*, (Dordrecht: Kluwer), p. 175

Sterken, C., & Staubermann, K. (2000). "Visual Magnitudes based on Zoellner's Catalogue," in C. Sterken & K. Staubermann (eds.), *Karl Friedrich Zollner and the Historical Dimension of Astronomical Photometry*, (Brussels: VUBPress), p. 95

Sterken, C., Snowden, M., Africano, J. et al. (1986). *Astronomy and Astrophysics Supplement Series*, **66**, 11

Sterken, C., et al. (2008). *Astronomy and Astrophysics*, **484**, 463

Stetson, H. T. (1916). *Astrophysical Journal*, **43**, 253–285

Stetson, H. T. (1916). *Astrophysical Journal*, **43**, 325–340

Stetson, H. T., & Carpenter, E. F. (1923). *Astrophysical Journal*, **58**, 36–45

Stoy, R. H. (1970). " The Initial Performance of the Galaxy Machine," in W. J. Luyten (ed.), *Proper Motions.*, Proc., IAU Colloq. 7, (Minneapolis: University of Minnesota), p. 48

Strömgren, B. (1937). In B. Strömgren (ed.), *Handbuch der Experimentalphysik*, Band **26**, *Astrophysik*, (Leipzig: Akademische Verlagsgesellschaft), p. 321

Strömgren, B. (1966). "Spectral Classification Through Photoelectric Narrow-Band Photometry," *Annual Review of Astronomy and Astrophysics*, **4**, 433

Toomer, G. J. (1984). *Ptolemy's Almagest*, (New York: Springer)

Tulenius, A. (1740). *Dissertatio astronomica de constellatione Arietis*, (Stockholm: Homiae)

Walker, A. R. (1993). "Photometry with CCDs," in C. J. Butler & I. Elliott (eds.), *Stellar Photometry*. (Cambridge: Cambridge University Press), p. 278

Weaver, H. F. (1946). *Popular Astronomy*, **54**, 211, et seq

Wesselink, A. J. (1941). *Leiden Annalen*, **17**(3)

Whitford, A. E. (1932). *Astrophysical Journal*, **76**, 213

Whitford, A. E. (1962). *Handbuch der Physik*, **54**, 240

Whitford, A. E., & Kron, G. E. (1937). *The Review of Scientific Instruments*, **8**, 78

Wisniewski, W. Z. (1993). Comment in C. J. Butler & I. Elliott (eds.), *Stellar Photometry*, (Cambridge: Cambridge University Press), p. 38

Young, A. T. (1967). *Monthly Notices of the Royal Astronomical Society*, **135**, 175

Young, A. T. (1974). "Observational Technique and Data Reduction," in N. Carleton (ed.), *Methods of Experimental Physics* Vol. **12**, Astrophysics, Part A: Optical and Infrared, Chap. 3, (New York: Academic), p. 105

Young, A. T. (1984). In W. J. Borucki & A. T. Young (eds.), *Proceedings, Workshop on Improvements to Photometry*. NASA Conf. Publ. 2350, 8

Young, A. T. (1992). "Improvements to Photometry V. High-order Moments in Transformation Theory," *Astronomy and Astrophysics*, **257**, 366–388

Young, A. T. (1993). "High-Precision Photometry," in C. J. Butler and I. Elliott (eds.), *Stellar Photometry – Current Techniques and Future Developments*, (Cambridge: University Press), p. 80

Young, A. T. (1994). In C. Sterken & M. de Groot (eds.), *The Impact of Long-term Monitoring on Variable Star Research: Astrophysics, Instrumentation, Data Handling, Archiving. NATO Advanced Science Institutes Series C: Mathematical and Physical Sciences*, (Dordrecht: Kluwer)

The High Road to Astronomical Photometric Precision: Differential Photometry

E.F. Milone and Jan Willem Pel

1 The Basics of Differences

One might think from the combination of errors that a difference between equally precise individual magnitudes would result in increased uncertainty, or decreased precision, by a factor of $\sqrt{2}$ in their difference. This is frequently the case in computing a color index from separate magnitudes; in fact some differential photometers have been created to observe only a single star in two or more passbands simultaneously in order to decrease the error in the measured color index. Nevertheless, ground-based observations can be improved by comparing the light of two stars in close proximity because of several conditions:

1. First-order atmospheric extinction (i.e., extinction independent of star color) depends on airmass (roughly $\propto$sec z), therefore if $\Delta z \approx 0$, so is this component of the general atmospheric extinction
2. First-order atmospheric extinction can vary both temporally and spacially; hence, simultaneous or nearly simultaneous observations of two stars in close proximity on the sky can improve the error budget under such conditions
3. Color-dependent extinction (sometimes called "second-order extinction") arises in part from atmospheric Rayleigh scattering but if two stars of similar intrinsic color index are observed, some effects may cancel
4. Effects of errors in transformation coefficients can be reduced somewhat by selection of stars of closely similar apparent spectral energy distributions (SEDs) if observed at low air mass
5. Variations in background flux and sky brightness can be minimized by differential photometry if sampled more quickly than the sky variations
6. When chopping can be applied, the noise at frequencies other than the chopping frequency can be minimized

E.F. Milone (✉)
RAO, Department of Physics and Astronomy, University of Calgary, 2500 University Drive, NW, Calgary, AB, Canada T2N 1N4
e-mail: milone@ucalgary.ca

E.F. Milone and C. Sterken (eds.), *Astronomical Photometry: Past, Present, and Future*, Astrophysics and Space Science Library 373, DOI 10.1007/978-1-4419-8050-2_2,

We demonstrate these points formally in the following discussion. The total measured power collected from a stellar source can be described in the expression;

$$\mathscr{P}_{\rm obs} = \frac{\pi}{4} \cdot D^2 \cdot \int_0^{\infty} \left[(\mathscr{F}_* + \mathscr{F}_{\rm bgd}) \cdot t_{\rm atm}\, t_{\rm tel} + \mathscr{F}_{\rm sky} \cdot t_{\rm atm'}\, t_{\rm tel'} + \sum_i^n \mathscr{F}_i^{\rm local} \cdot t_i \right] q_\lambda \, d\lambda, \tag{1}$$

where $\mathscr{F}$ is the flux from the particular source indicated by subscript; t is the transmission coefficient, so that $t_{\rm atm}$ is the transmission through the atmosphere, and $t_{\rm tel}$ through the telescope optics including filter(s) and windows in the optical path, and t_i, transmissions of the various local fluxes; primed transmission quantitities are associated with sky brightness (generally these need not be identical to the corresponding stellar or local values); D is the effective diameter of the light-gathering area of the telescope; q_λ is the wavelength-dependent quantum efficiency of the detector, and λ is wavelength. The integration is taken across the passband, but note that here the explicit passband function is defined by the optical transmission functions. The transmission factors and detector spectral sensitivity are sometimes explicitly combined into an instrumental passband function, S_λ. Background flux is that in the star field itself due to other stars as well as diffuse objects such as galaxies and nebulae. The sky flux includes moonlight, twilight, airglow, and aurora, but the treatment of these diffuse sources can be challenging because the light path may not be identical to that of light from a targeted object. The sources of local flux are the lights within and outside the telescope enclosure. We may assume that the observer can control these nuisance sources satisfactorily so we need not discuss them further. The collected power is readily converted into the number of detected photons per unit time with the expression,

$$n_{\rm meas} = \mathscr{P}_{\rm obs} / \left(hc\lambda_{\rm eff}^{-1}\right) + n_{\rm det}, \tag{2}$$

where the factor $(hc/\lambda_{\rm eff})^{-1}$ converts units of power (energy/time) into a photon count rate, and where the quantity λ_{eff} applies to the passband; a more exact treatment would involve division of (1) by the monochromatic factor ($hc\lambda^{-1}$) and integrated across the passband. n_{det} is an unwanted contribution from the detector or its associated electronics, again in counts per second. In photoelectric detectors, n_{det} arises from thermal emission within the detector and can be minimized by cooling, either by a cryogen such as dry ice (solid CO_2) or thermoelectrically. In a charge-coupled device (*CCD*) it arises partly from thermal emission, which similarly can be controlled by cooling, and partly from "read-out" detector noise, which can be measured. Read-out noise is usually small in current-day CCDs but may be important under such conditions as very low signal and background flux levels or very short exposure times. We will assume that this contribution, too, is manageable so it need not be discussed further. With (2), the time-integrated number of detected events can be written:

$$\mathscr{N}_{<\lambda>} = \frac{1}{4}\pi D^2 \left[<\mathscr{F}_{*+\rm bkg}> \cdot <t_{\rm atm}> \cdot <t_{\rm tel}> + <\mathscr{F}_{\rm sky}> \cdot <t_{\rm atm'}> \cdot <t_{\rm tel'}> \right]$$
$$\cdot <q_{<\lambda>}> \cdot \Delta\lambda \cdot \tau \cdot (hc/\lambda_{\rm eff})^{-1}, \tag{3}$$

where τ is the integration/exposure time of the observation, and the angle brackets $(< >)$ indicate averages over τ, or the effective passband width, $\Delta\lambda$. Hopefully most, if not all, of the background flux arrives at the detector through the same optical path as the target flux, although (depending on the design of the optical system) the sky flux may not. The difference in pathways may be important for glass and interference filters because of the dependence of λ on angle of incidence. In principle, the background and sky fluxes can be subtracted once they are measured, although their contributions to the noise remain.

Equation (3) is expressible in terms of magnitude measure but remains inconvenient unless the background and sky terms are subtracted first.

Equations (2) and (3) have useful applications, but it suffices for present purposes to proceed with (1). Over an integration time, τ, we arrive at the radiant energy collected in the interval:

$$\mathscr{E} = \mathscr{P} \cdot \tau. \tag{4}$$

Differential photometry can improve matters by eliminating some sources of noise and by minimizing others, but the detailed treatment determines how effective the differential photometry can be. If we express the passband brightness of one star relative to another by substituting from (1) and then differencing (4), without first subtracting the sky, we are left with several inconvenient terms:

$$\begin{aligned} \mathscr{E}_1 - \mathscr{E}_2 = \frac{\pi}{4} \cdot D^2 \cdot \{ & [(< \mathscr{F}_{*,1} > - < \mathscr{F}_{*,2} >) \cdot < t_{\text{atm}} > \cdot < t_{\text{tel}} >] \\ & + [(< \mathscr{F}_{\text{sky},1} > - < \mathscr{F}_{\text{sky},2} >) \cdot < t_{\text{atm}'} > \cdot < t_{\text{tel}'} >]\} \cdot < q > \cdot \tau \cdot \Delta\lambda, \end{aligned} \tag{5}$$

where averages are taken over the integration time interval as well as over the effective spectral bandwidth of the passband, and the background fluxes are assumed to be identical. If the sky fluxes are also identical, the second bracketed term vanishes; if they are not, or if the backgrounds are not identical, lack of detailed investigation of these terms will impact both precision and accuracy. If, on the other hand, the sky is carefully measured and subtracted from the observed radiant energy, so that

$$\mathscr{E}_{\text{target}} - \mathscr{E}_{\text{sky}} = \frac{\pi}{4} \cdot D^2 \cdot < \mathscr{F}_{\text{target}} > \cdot < t_{\text{atm}} > \cdot < t_{\text{tel}} > \cdot q \cdot \tau \cdot \Delta\lambda, \tag{6}$$

for each of two stars, the ratio of the sky-subtracted detected energies becomes

$$\frac{\mathscr{E}_{*1} - \mathscr{E}_{\text{sky},1}}{\mathscr{E}_{*2} - \mathscr{E}_{\text{sky},2}} = \frac{\mathscr{F}_{*1}}{\mathscr{F}_{*2}} \cdot \frac{< t_{\text{atm},*1} >}{< t_{\text{atm},*2} >}, \tag{7}$$

after common terms are divided out. From this, a differential magnitude can be obtained,

$$m_1 - m_2 = -2.5 \cdot log \left\{ \frac{\mathscr{F}_{*1}}{\mathscr{F}_{*2}} \right\} - 2.5 \cdot log \left\{ \frac{< t_{\text{atm},1} >}{< t_{\text{atm},2} >} \right\}. \tag{8}$$

Johnson (1962) states that determining simultaneous sky brightness can produce an improvement of 20–100% in the resulting photometric accuracy depending on the amount of sky variation. Shortly we will consider the sources of noise that can afflict such differential measurements; but first we need to consider the sources of error in photometric measurements generally. These are:

- Scintillation noise
- Target photon noise (sometimes called "shot noise")
- Sky photon noise
- Thermal environment noise
- Detector noise; and, ultimately
- Transformation errors

Scintillation noise is proportional to the signal, increases with airmass, and is dominant for bright stars, but can be minimized by observing with larger telescopes that gather light through a wider column of atmosphere; the dependence varies with the inverse square of telescope diameter. Typical scintillation amplitudes for a 30-cm telescope are about 1% with frequencies of ~1–100 Hz. The signal to noise ratio, *SNR*, for this source is, moreover, independent of signal. Code and Liller (1962) compare the noise contributed by "seeing" with other noise sources over the range of apparent magnitudes $0 \leq V \leq 20$.

Photons follow Poisson statistics, so that $SNR \propto \sqrt{\mathcal{N}}$, where $\mathcal{N}$ is the number of recorded events, as in (3). Because both target and sky shot noise suffer these fluctuations, the noise enters from both measurements. In addition, sky brightness may vary greatly due to scudding clouds before the Moon, aurorae, and merely sky glow. Yes we said "scudding clouds before the moon" because photometric skies are not always photometric from zenith to horizon, as any observer can attest. Thus both the intrinsic variations as well as the shot noise connected with these varying sources of spurious light may contribute to the noise.

The treatment of emission from the thermal environment and noise is important for infrared astronomy, and is usually treated by cryogenic cooling of the detector with liquid nitrogen, or, for the thermal infrared, beyond 2 μm, with liquid helium. The additional sources of impairment of infrared photometry, along with their remedies, are described in the Milone and Young paper in this volume and in the citations of that paper. Instrumental noise these days arises mainly from the detector; amplifier noise, once highly significant in photoelectric photometry, is now relatively insignificant. Although read-noise has been a significant source of noise in CCD observations, as Howell's paper in this volume shows, this source of error has decreased greatly in modern CCD photometry. For completeness, we mention also noise due to transformation error. This is not a source of random error, but of systematic error because it depends on the spectral energy distribution and other factors. It may, however, be rendered less important by careful selection of comparison star. A fuller discussion of errors in differential photometry can be found in Young et al. (1991), who conclude that even with careful selection of comparison stars, manual alternate observations of target and comparison stars can rarely produce precision better than 0.5%. Automated photometry, however, can do better sometimes, as we demonstrate below.

In general, the combination of random noise sources gives the net noise level:

$$\mathcal{N}_{\text{total}} = \sqrt{\left[\mathcal{N}^2_{\text{scint}} + \mathcal{N}^2_{\text{shot}} + \mathcal{N}^2_{\text{sky}} + \mathcal{N}^2_{\text{det}}\right]}. \tag{9}$$

The combination of errors can be applied to (7) in order to find the precision of a differential measurement:

$$e_r = \left\{\left[e^2_{*1} \cdot \left(\frac{t_{\text{a}1}}{\mathcal{F}_{*2}t_{\text{a}2}}\right)^2\right] + \left[e^2_{*2} \cdot \left(\frac{\mathcal{F}_{*1}t_{\text{a}1}}{\mathcal{F}^2_{*2}t_{\text{a}2}}\right)^2\right] + \left[e^2_{t_{\text{a}1}} \cdot \left(\frac{\mathcal{F}_{*1}}{\mathcal{F}_{*2}t_{\text{a}2}}\right)^2\right] + \left[e^2_{t_{\text{a}2}} \cdot \left(\frac{\mathcal{F}_{*1}t_{\text{a}1}}{\mathcal{F}_{*2}t^2_{\text{a}2}}\right)^2\right]\right\}^{0.5}, \tag{10}$$

but we note that $e_{*1,2}$ include sky and background noise contributions.

Now we ask in what ways differential photometry can improve both accuracy and precision. The selection of a non-variable comparison star has much to do with it. By "non-variable" we mean a star found to be not significantly variable at the desired precision level over at least the time interval of the observing run. Three general criteria are followed:

1. Small angular distance from the variable.
   This criterion helps to assure a small atmospheric extinction difference between the stars
2. Greater brightness than the variable (with some exceptions).
   This can improve photon statistics. There are occasions, however, where systematic effects may occur if the comparison star is substantially brighter than the variable, or vice-versa. Normally one would not select, except of necessity, a comparison star that is much fainter than the variable
3. Similarity in apparent SED to the variable.
   Color indices may be more useful than spectral type matching if there is substantial difference in interstellar reddening between the two stars, because such an effect redistributes the spectral energy distribution, effectively changing the effective wavelength of the passband, which leads to a significant color term difference. A second-order effect may remain, however, due to the shift in λ_{eff} introduced by differences in spectral features within the passband. It is the distribution of flux across the passband that gives rise to the color term (sometimes incorrectly referred to as the "second-order" term) in extinction. We discuss this color dependence further, below

Sometimes it is desirable to add a fourth criterion (Milone 1967): Equality in zenith distance at some selected hour angle of the pair so that the differential air mass is near zero when needed. Usually comparison stars are not so numerous that one can fulfil all of these criteria, but in crowded regions, this may be possible. This also may be useful, however, where the best comparison stars are not within a few arcmins of the target star. The required position angle, p, and optimized hour angle can be derived from the equation for the differential airmass. Ignoring corrections for atmospheric curvature and refraction, we obtain

$$dX = -X^2 \cdot (\cos\delta \sin\phi \, d\delta - \sin\delta \cos\phi \cos H \, d\delta - \cos\delta \cos\phi \sin H \, dH), \tag{11}$$

where $X = \sec z$, δ = mean declination, ϕ = site latitude, and H = mean hour angle. The nulling of this quantity yields

$$dH/d\delta = -d\alpha/d\delta = \frac{\tan\phi - \tan\delta \cdot \cos H}{\sin H}, \tag{12}$$

where, for simultaneous observations, $\mathrm{dH} = -\mathrm{d}\alpha$, where $\mathrm{d}\alpha$ is the difference in right ascension between the two stars.

The position angle can be expressed as

$$\tan p = \frac{d\alpha \cdot \cos\delta}{d\delta}, \tag{13}$$

so that

$$\tan p = -\frac{(\tan\phi - \tan\delta \cdot \cos H) \cdot \cos\delta}{\sin H}. \tag{14}$$

The optimized hour angle can be found from the solution of (12):

$$\cos H = + \frac{\tan\phi \cdot \tan\delta \pm (d\alpha/d\delta) \cdot \left[\left(\tan^2\delta - \tan^2\phi + (d\alpha/d\delta)^2\right)\right]^{0.5}}{(d\alpha/d\delta)^2 + \tan^2\phi}. \tag{15}$$

Given a comparison star that meets the first three criteria, it is an easy matter to use (15), to determine what part of a night provides a minimal airmass difference. A fuller exposition is given in (Milone 1967).

It should be noted that a full characterization of the extinction on any given night is very difficult to achieve with broadband filter systems. Young (1988) suggests that such a determination requires overlapping passbands which can satisfy the sampling theorem, something the *UBV* system alone cannot do; on the other hand, narrower overlapping passbands, such as the Geneva system B_1 and B_2 undersample the spectrum in other ways. Even if two stars are identical in intrinsic color (and spectral type), the slight changes caused by color extinction will results in departures from millimagnitude extinction in the *B* band. Thanks to the Chappuis bands of ozone, however, there is a very much reduced atmospheric reddening effect within the *V* passband. This makes highly precise differential photometry possible if still difficult. To cite Young et al. (1991) "... if one is willing to live with fairly exacting contraints" in terms of comparison star selection, air mass, and magnitude and color extinction measurements, "then millimagnitude differential photometry of some stars may be possible with existing systems such as the Johnson-Morgan B and V."

2 Differential Visual Photometry

The idea to compare visually the brightness of one object with another exploits the excellent ability of the human eye to do just this. Probably the earliest such documented use of this principle in astronomy was by Pierre Bouguer in 1725 (Bouguer 1729), of "Bouguer extinction" renown, who used a standard candle to illuminate a surface and compare the illumination with that produced by the Sun and Moon. John Herschel made use of a demagnified image of the Moon to measure stellar brightnesses of Southern stars in 1836. About this time, C. A. Steinheil (1837) developed his prism photometer in conjunction with a refracting telescope, the objective lens of which was split in two. Each half of this lens had the same focal length, but could be independently focused. Consequently, a star imaged with one objective component could be compared with another star imaged by the other objective component (a prism in front of one component introduced a second star from outside the field of view). By separating the two objectives so that the stars' extrafocal (spread-out) images had the same intensity, Steinheil (1837) created a kind of null photometer; the difference in distance of the two objectives from the focal plane provided a measure of the relative stellar brightnesses.

In 1850, F. Arago (1858) suggested that the principle of polarization could be used to measure starlight. This seminal idea led to the most precise and accurate means of determining relative brightness available to visual photometry (see Bastien's paper in this volume). J. C. F. Zöllner (see Hermann (2002) for a recent biography and portions of his work) followed Arago's suggestion to create the first polarizing comparative photometer, with which a focused star's brightness could be compared with that of an artifical star. The light of the artifical star (a kerosene lamp) was polarized by passage through a fixed Nicol prism and then through a rotating analyzer; this allowed the light to be dimmed to match the apparent brightness of a star. As part of a submission for a photometry prize offered by the Academy of Sciences of Vienna, he included the design and a list of 226 stars observed with it. Although Zöllner was not awarded the prize, his work was subsequently published (Zöllner 1861) and proved very influential. Copies of Zöllner's instrument were used subsequently by a number of astronomers: C. Pierce at Harvard (from $\sim$1871), J. Wolff at Palermo and Bonn (from 1869), G. Müller at Potsdam (from 1877) who with P. Kempf produced the *Potsdamer Durchmusterung* survey, and E. Lindemann at Pulkovo (from 1870). The highest precision attainable with these early comparative devices was not great, not better than $\sim$5%, and typically worse.

In the summer of 1877, following in the footsteps of Bond and Pierce, E. C. Pickering (1879) started a new program to carry out visual differential photometry at Harvard, and experimented with a variety of apparatus to develop the best instruments for this purpose. The work was begun with the Zöllner astrophotometer but this was abandoned because the artificial star was considered unsuitable. Pickering and his associates (O. Wendell and A. Searle) found it insufficiently constant, and, moreover, the image could not be reduced to a small enough size to match the stellar seeing disk. There was a certain degree of irony in this; years earlier Zöllner had criticized G. P. Bond (1861) for his use of a highly variable flame that was said to match the color of the moon, in order to obtain differential measurements of the Sun and Moon with the flame, and thus to each other. Pickering's group then tried to use various images of bright stars, dimmed, as had been the artificial star, by a double-edged prism, used in conjunction with a rotatable Nicol prism. The latter acted as a polarization analyzer, and, as it rotated, reduced the light of a star. Most of the instrumental configurations contained two telescopes, with the dimming performed on light through the smaller of the two, which was used for the brighter star. Figure 1 shows one of the earlier two-star instruments in use at Harvard (Pickering 1879, Figs. 5 and 6, p. 7). Instrument "I" used an aperture stop rather than a polarizer to vary light from the brighter star.

Decades later a version of the Pickering polarizing photometer, as the visual two-star instrument was then called, was still in use at Harvard (Wendell 1909) and a copy was being used at Princeton (Dugan 1911). Indeed, the latter was used at Princeton through the 1930s. John E. Merrill reported (private communication to EFM, 1980) that the observations of RW Com that he obtained with the instrument (reported by Russell (1939) and published in Milone et al. (1980)), had a typical mean photometric precision of ~2%. This was about the maximum precision obtainable with such an instrument.

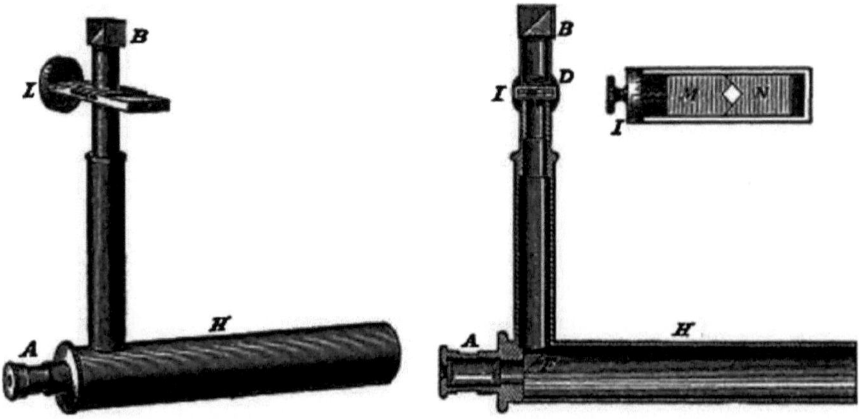

Fig. 1 A version of the visual two–star photometers in use at Harvard in the nineteenth and early twentieth century century. "Photometer I", as illustrated and described by Pickering Pickering (1879) did not include polarizing elements but rather a variable aperture in one of the two telescopes to vary the brightness of one of the two stars. Legend: *A*, eyepiece in larger telescope to view the fainter star; *B*, prism; *D*, objective for the smaller telescope; *F*, prism (near *A* within telescope tube *H*) to bring image of brighter star into field of view; *M*, *N*, two thin slabs of brass with V-shaped ends, forming a square "cats-eye" opening, adjustable with calibrated knob, *I* (image: courtesy Harvard College Observatory)

3 Differential Photographic Photometry

Although photography was invented in 1837 and first astronomical images were obtained in 1840 by J. W. Draper, only after the invention of dry gelatine emulsions in 1870 did the photographic plate become the primary astronomical detector. The differential visual comparisons made of early type stars near the North Celestial Pole, that had been pioneered by Harvard astronomers, were further extended to the photographic realm, and the use of a visual dye led to the development of multi-wavelength photometry with photographic (m_{pg}) and photovisual (m_{pv}) magnitudes.

As noted in the article by Sterken et al. in the present volume, differential photographic photometry provided the highest possible precision for photographic work; this is evidenced by the work of Jordan (1923) on the short-period eclipsing binary star system RW Com, by Plaut (1940) on WW Dra, the eclipsing brighter component of the wide double Σ2092, and by Wesselink (1941) on SZ Cam, the eclipsing binary component of the visual binary Σ485. Wesselink found that the difference in the $m_{pg} - m_{pv}$ color index of the visual pair was only 0.05. On the plane of the sky they are only 17.9 arcsec apart. The brightness of the comparison star is about equal to the average brightness of SZ Cam, which is the northern component of this pair. Thus the companion satisfies most of the criteria mentioned in Sect. 1. Wesselink showed that any variation of the southern component was within ±0.02 magn., so it was demonstrated to be a suitable comparison star (one can be unlucky in choice of comparison and check stars if these have not been studied previously; Wesselink was fortunate here). To improve the precision of the differential photometry, Wesselink went further than merely having at hand a close and well-matched comparison star. He made use of a diffraction grating placed above the objective of a long-focal length refractor. The 29 cm objective grating produced first order diffraction images removed from the zero-order images by 12 arcsec (corresponding to 0.29 mm on the plates) with known magnitude difference between the zero- and first-order images. A typical procedure was to take 11 (or more) exposures per plate, resulting in 44 stellar images. Wesselink measured the images with the Schilt photometer (see Sect. 3 of the Sterken et al. paper in this volume for a discussion of these instruments). He examined carefully the difference in the constant dispersion produced by the objective grating and the atmospheric dispersion, that varies in both direction and extent, with hour angle; by selecting the dispersions only at 45 and 225°, so that the dispersion did not overlap any star images, and by using a comparison star of similar brightness to the variable, he was able to reduce the effects of systematic error. He found in the end no systematic error in Δm depending on the hour angle. The maximum mean precision attained is ~2%.

4 Differential Photoelectric Photometry

Accurate differential photoelectric photometry became possible only after the appearance of photomultiplier tubes. Although the first photomultipliers were developed already in the mid-1930s, in particular at the RCA laboratories, it took some time before these new devices became widely used in astronomical photometry. This was primarily due to World War II. After the war, the superior properties of photomultipliers were quickly recognized by the astronomical community and many observatories started to develop photometric instruments based on the first generation of commercially available photomultipliers, the RCA 931 and 1P21 type tubes. One of the pioneers in this new field of astronomical instrumentation was Theodore Walraven (1916–2008), whose passing occurred just as the present volume was being planned. In recognition, we will devote this section to his differential photometry work and legacy.

Immediately after his appointment to the Leiden Observatory in November 1946, and while still working on his Ph.D. thesis, Walraven started experimenting with photomultipliers. By the end of March 1947, always assisted by his wife Johanna, he made his first photoelectric observations with an uncooled 1P21 tube on the 48-cm Zunderman reflector (Fig. 2). Extensive series of observations were obtained of RR Lyrae, which resulted eventually in Walraven's classical analysis of the "Blazhko-effect" in RR Lyrae, (Walraven 1949) rediscussed later in the landmark review paper together with Paul Ledoux in Handbuch der Physik (Ledoux and Walraven 1956).

These early photoelectric measurements on rapidly changing variables such as RR Lyr had a decisive influence on Walraven's designs of subsequent photometric instruments. The accuracy that could be achieved with single-channel photometers in the light- and colour-curves of short-period variables was clearly limited by extinction variations. Obviously big improvements could be obtained by

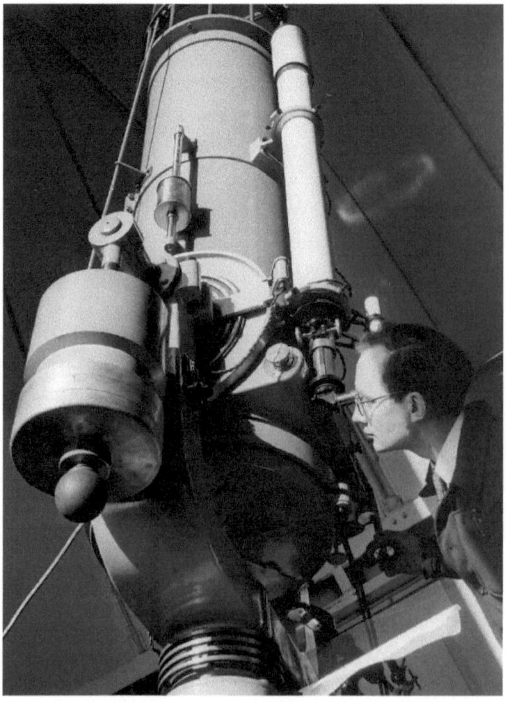

Fig. 2 Theodore Walraven at the 48-cm Zunderman reflector of Leiden Observatory in 1947 (photo: courtesy Leiden Observatory)

applying differential techniques. While the construction of several single-channel photometers (both cooled and uncooled) for the telescopes in Leiden and for the Leiden Southern Station in Johannesburg was going on, the first steps towards differential photometry were made. This was done in two directions: simultaneous measurement of two stars – variable and comparison star – in a given filter passband, and simultaneous measurement in several passbands of one star. Surprisingly, the more challenging of these two techniques, simultaneous multi-passband photometry, had been tackled already in 1949 with the construction of a prototype multi-channel photometer for simultaneous observations in four filters. This development would result eventually in the well-known Walraven five-channel photometer (see below).

4.1 The Walraven Experiments

In parallel with the construction of the other photometers, a "differential nulling photometer" was designed by Walraven in 1950 with the specific aim to observe short-period variables (quasi)-simultaneously with a nearby reference star. Although publication of a full description of this instrument was announced as forthcoming by Walraven on several occasions, this unfortunately never happened. On the basis of the available information (Walraven 1952a,b, 1953, 1955) the principle of the instrument can be summarized as follows. In the first version of the differential photometer the light beams from both stars were directed alternatingly to the same phototube in order to eliminate the problem of sensitivity variations between two phototubes. This light switching was done by means of an oscillating double diaphragm in the telescope focal plane and a set of optics that produced pupil images for both beams on the same spot of the RCA 931A photocathode. The phototube output thus contained an AC component with amplitude proportional to the intensity difference between the two stars. By means of a neutral density wedge in one of the two light paths this intensity difference could be nulled. The setting of this wedge was done automatically by a DC motor in a servo loop driven by the AC amplitude of the phototube output signal. The actual measurement consisted in automatic and continuous registration of the wedge position on a chart recorder. The only necessary human interventions during the observations were checks on the telescope guiding, typically once per 30 min. Due to the nature of the optical transmission of the wedge the nulling-position curve on the recorder multiplied by a scale factor yielded a direct light curve registration in magnitudes, which greatly reduced the photometric reduction effort and allowed on-line determination of light curve parameters. The wedge scale was calibrated by observing the comparison star for different wedge settings in "DC-mode." Figure 3, kindly made available by Adrian A. Disco, who was involved in the construction of this early prototype instrument, shows its installation on the 33-cm refractor of Leiden Observatory in July, 1950. According to Disco, the instrument was tested on the star XZ Cygni.

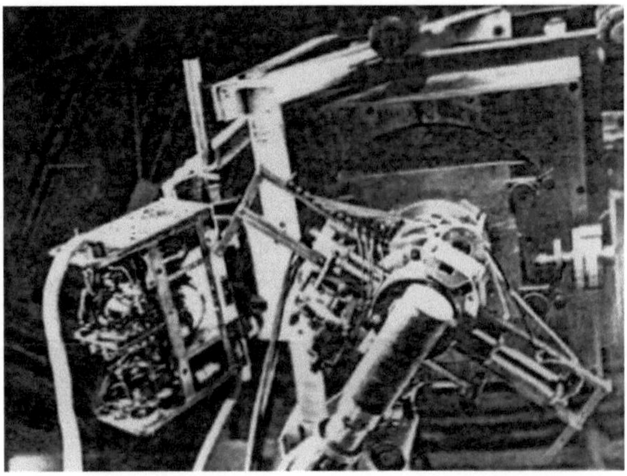

Fig. 3 Prototype of the differential wedge photometer (one cell version) during tests on the 33-cm refractor of the Leiden Observatory, in July, 1950. Courtesy, Adrian A. Disco, who worked on the servo-mechanism

After several modifications of the prototype instrument, including experiments with photon-counting, the Walravens moved to the Leiden Southern Station in Johannesburg, where the new photometer was mounted on one of the twin 40 cm refractors of the "Rockefeller" telescope in March 1951. This equipment was then used for extensive observations of the multi-period variable AI Vel (then called "RR Lyrae–type" variable, but nowadays classified as high-amplitude member of the Delta Scuti family). In the meantime the work on improvements of the photometer continued, resulting in a two-channel version. Once the improved two-channel wedge photometer was completed in the course of 1951 the measurements were resumed with that instrument and AI Vel and SX Phe were intensively monitored until April 1953. Examples of the superb light curves produced by the nulling technique are given in Fig. 4 (from Walraven (1955)). Walraven's papers on the pulsations of AI Vel, SX Phe and RR Lyr (Walraven 1952a,b, 1955) still stand as classics in the variable stars literature.

The two-channel differential photometer used the same optical wedge nulling technique, but this time there was a set of four fixed input diaphragms – two for the two stars and two for nearby sky positions – and the more elaborate optics included pairs of reflecting prisms for centering of the stars in the diaphrams and and two prism sets for beam switching. A brief description of the two-channel instrument was given by Walraven in his paper presented (and read, by Bengt Strömgren) at the small photometry symposium in Philadelphia on December 31, 1951 (Walraven 1953). A schematic layout of the switching scheme is given in Fig. 5.

After the return of the Walraven family to Leiden in 1953 the observations with the differential photometer stopped and the instrument was dismantled. However, Walraven never lost interest in this elegant instrument or in his favourite star. A quarter of a century later, and back in South Africa, the Walravens built a new version of the differential photometer using the original optical wedge and started re-observing

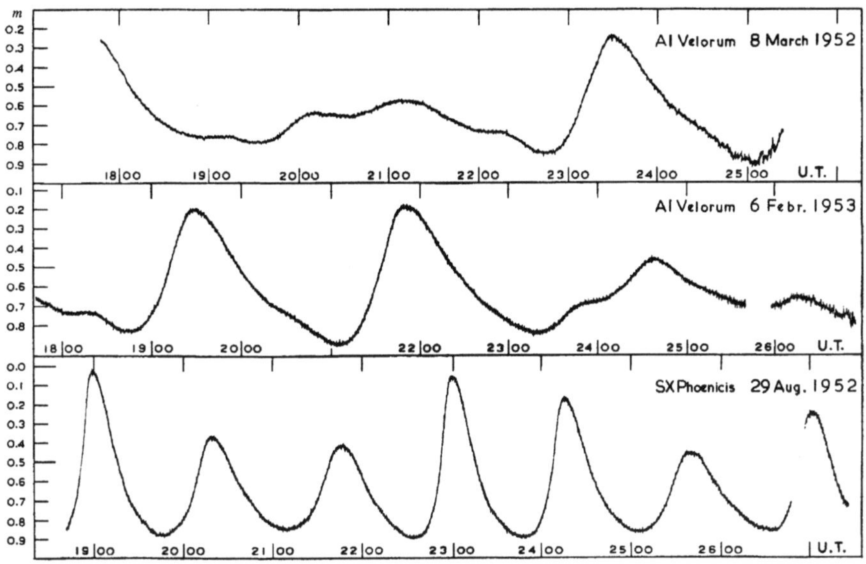

Fig. 4 A sample of the light curves of AI Velorum and SX Phoenicis, obtained with the 2-channel differential wedge photometer and the 40-cm "Rockefeller" refractor of the Leiden Southern Station at Union Observatory, Johannesburg, South Africa (from Walraven (1955))

AI Vel in 1979. After retirement in 1980, when the Walravens moved to the small town of Cornelia in the Orange Free State, these observations were continued with their private, fully automated 40-cm telescope and the novel simultaneous six-colour photometer that Walraven had developed during the last part of his career. The combined dataset of AI Vel photometry spanning a period of 38 years was eventually analyzed in collaboration with Louis Balona (Walraven et al. 1992). This impressive study revealed additional pulsation modes and systematic period ratio changes.

Back in Leiden in 1953, Walraven had taken up the work on the other differential photometry approach: simultaneous photometry in multiple wavelength bands. The prototype four-channel photometer mentioned above eventually evolved into the five-channel (*VBLUW*) photometer that was built for use on the new 91-cm "Lightcollector" reflector. While this telescope was being manufactured by the Rademakers company in Rotterdam, the photometer was developed at Leiden Observatory during 1954–1957.

The *Lightcollector* and its photometer were specifically designed as an integrated system for efficient photometric observations on individual stars (cf. Fig. 6; Walraven and Walraven 1960; Pel and Lub 2007). The telescope had only a small field with good image quality, but it was equipped with fast and accurate electronic presetting. The photometer was based on an ingenious quartz/calcite crystal optics filter that produced five well-defined and ultra-stable passbands in the visual/near-UV. These passbands were separated in a 2-prism spectrograph and measured simultaneously by five 1P21/1P28 photocells and current-integrating amplifiers.

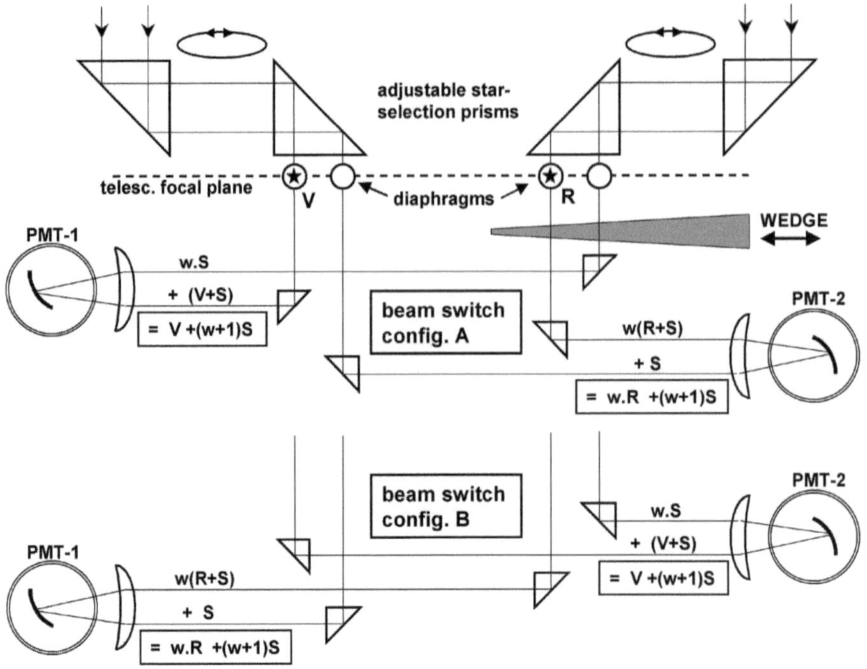

Fig. 5 Schematic layout of the beam switching optics in Walraven's 2-channel differential photometer. The two stars (variable "V" and reference "R") are centered into two of the four identical diaphragms in the telescope focal plane by means of the star-selection prisms. The other two diaphragms select sky background, "s". One pair of beams passes through a neutral density wedge that can be moved by a DC-motor. Two alternating sets of four beam-switching prisms direct the beams to the photomultiplier tubes PMT-1 and PMT-2. By switching between the beam configurations "A" and "B" both phototube signals are modulated in anti-phase with the same amplitude (V - w.R), where w refers to the wedge attenuation factor, eliminating the sky background. This amplitude drives the servocontrol loop that actuates the optical wedge until (V - w.R) = 0. The signal modulations in both phototubes are then nulled simultaneously and the wedge position gives the magnitude difference $V - R$

All photometer optics were of UV-transmitting materials (quartz/calcite/fused silica) and, as in all of Walraven's instruments, had been made mostly by Johanna Walraven. With two bands (V,B) longward of the Balmer jump, two bands (U,W) in the Balmer continuum and the L band at the confluence of the higher Balmer lines, the ($VBLUW$) photometry was designed as a photoelectric counterpart of the three-dimensional classification of (photographic) stellar spectra by Barbier, Chalonge and Divan, with which Walraven had become closely familiar during long observing runs at the Observatoire de Haute Provence in 1949–1950.

In 1958 the 91-cm telescope was erected at the new location of the Leiden Southern Station, the Union Observatory Annexe at Hartebeespoortdam. Routine observations with the new telescope and photometer started in 1959 with a massive observing program on OB-stars. For many years the observations remained focused

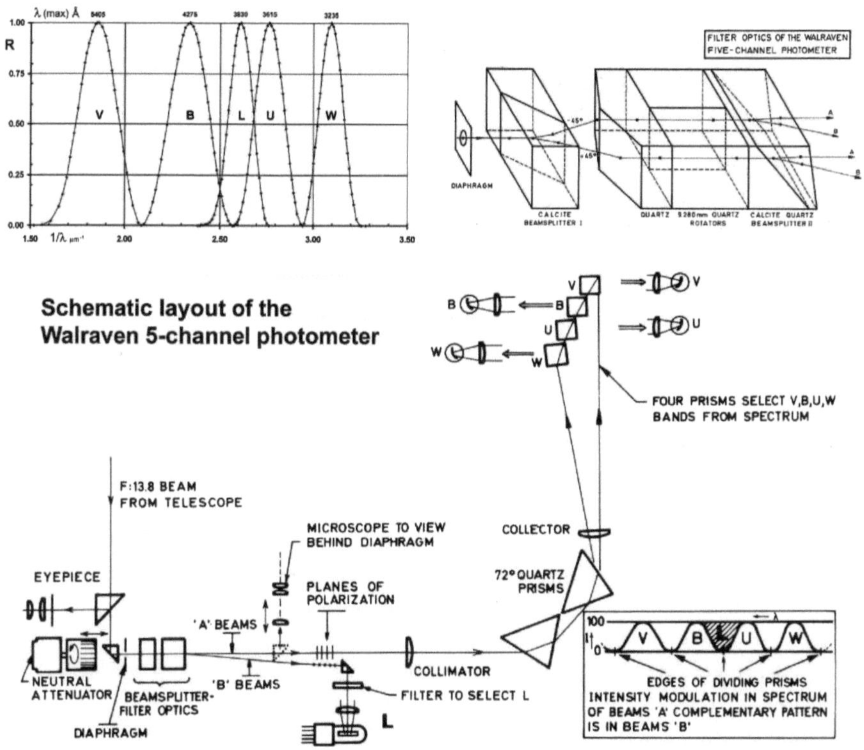

Fig. 6 Overview and schematic layout of the Walraven *VBLUW* photometer (from Pel and Lub 2007)

on early-type stars, but when the (*VBLUW*) system turned out to be very effective also for the determination of ($T_{\rm eff}$, $log\ g$, $[Fe/H]$) of A-F-G stars the scope widened to later spectral types and big programs on Cepheids and RR Lyrae stars followed. A more detailed discussion of the 32 years of productive life of the Walraven five-channel photometer (Hartebeespoortdam 1959–1978, ESO La Silla 1979–1991) is outside the scope of this paper, but some final remarks about the application of differential principles with this instrument are in order.

Already during his first photometric observations Walraven had been struck by the fact that atmospheric extinction variations at different wavelengths were strongly correlated. This implied that the accuracy of colour measurements could be improved greatly by simultaneous measurements in multiple passbands and it explained why Walraven focused so strongly on multichannel instruments and differential techniques. The (*VBLUW*) photometer proved the point very convincingly. Whereas in single-channel photometry the errors in colours were usually larger than the individual filter magnitudes, merely due to the combination of errors, (*VBLUW*) color indices were 2–3 times more accurate than the individual magnitudes, even for the $(U-W)$ colour, with band centers at 360 and 325 nm.

A differential approach was also followed in the calibration of the system of ($VBLUW$) standard stars. When the ($VBLUW$) photometry was first set up in 1958, a standard star system had to be built up 'from scratch'. This was done following a simple and elegant differential method with the following steps:

1. Selection of ~20 suitable standard star candidates, well distributed around the declination circle with DEC=(site geographic latitude): the "ring standards"
2. Frequent observations of pairs of ring standards in quick succession when Δ(airmass) ≈ 0. This yields per pair accurate differential magnitudes in all channels because instrumental gain drifts drop out (Δt very small) and a rough approximation of the extinction coefficients is sufficient for accurate differential extinction correction [$\Delta(sec\ z)$ very small]. In about 6 months all ring standard pairs are observed many times, covering the full 24 h circle, and the mean magnitude differences per pair are known very accurately
3. A least squares "ring closure" solution for the whole network of pairs and careful check on possible systematic effects (as function of time, position, brightness, colour) in the residuals. This fixes all differential magnitudes in all channels very precisely, but leaves the zero-points of the magnitude scales still undetermined
4. Definition of the zero-points, in this case done by choosing HD 144470 as primary ring standard with all four colour indices exactly zero and $V_{Walraven} \equiv V_{UBV_{Johnson}}$
5. The set of ring standards is now completely fixed, so extinction coefficients and instrumental zero-points for all observing nights can be determined
6. Secondary standards can be connected to the ring standards and final (extinction, zero-point)–solutions can be made for all nights with the full set of standard stars

During the 32 years of operation of the ($VBLUW$) photometer this procedure was repeated a few times, most extensively after the move of telescope+instrument to La Silla in 1979. In the latter case the residuals of the least-squares ring solution were extremely small: 0.00045 mag for V and [0.00025, 0.00031, 0.00031, 0.00027 mag] for [$(V-B)$,$(B-L)$,$(B-U)$,$(U-W)$] (Pel 1991; Pel and Lub 2007). This demonstrated beautifully the power of differential measurements. It should be noted that these very small residuals proved only that the internal consistency of the system was very high; systematic errors could still be present. However, extensive cross-checks with photometry in other systems, both ground-based (Johnson, Geneva and Strömgren photometry) and space-based (Hipparcos photometry), showed that the ($VBLUW$) system was truly free of systematics down to the level of 0.001 mag, a beautiful confirmation of the strength of Walraven's differential strategies.

4.2 Development of Other Photoelectric Differential Photometers

In the late 1950s and 1960s, other dual-channel photometers were designed and built at Kitt Peak National Observatory and elsewhere (Crawford 1958; Wood and

Lockwood 1967), the purpose of which was to observe in two passbands simultaneously. Thus the main purpose was to obtain color information.

Baum et al. (1959) describe a six-channel differential photometer which consisted of separate pickoffs for a star and nearby sky and funneled the light from each through separate filters to U, B, and V filters, but which is criticized by Johnson (1962) because of the unsatisfactory U filter available at that time, and because of the use of dichroic filters which would make transformations to the Johnson-Morgan UBV system, as he called it, difficult. The criticism is interesting because Johnson was one of the coauthors on the 6-channel photometer paper.

Beginning in the 1960s, relatively simple photometers were constructed that conveniently and systematically sampled two regions of sky, but not simultaneously. Such instruments were in use at Dyer Observatory (see Fig. 7) in the 1960s and other photometers with sky offsets were described by Visvanathan (1972) and by Kinman and Mahaffey (1974). We will not discuss the pioneering work on differential pulse-counting photometry by W. Blitzstein, resulting in PBPhot, because the latter is discussed extensively by Ambruster et al. in this volume.

An automated double-beam photometer was designed and constructed by McCord (1968). This photometer involved a mirror chopper that passed beams from different parts of the field alternately into the single detector. The chopping frequency was 30 Hz. A mechanical slide mechanism allowed a mirror to shift across the field of view, with a range of 10 arcmin. A similar chopping system but with

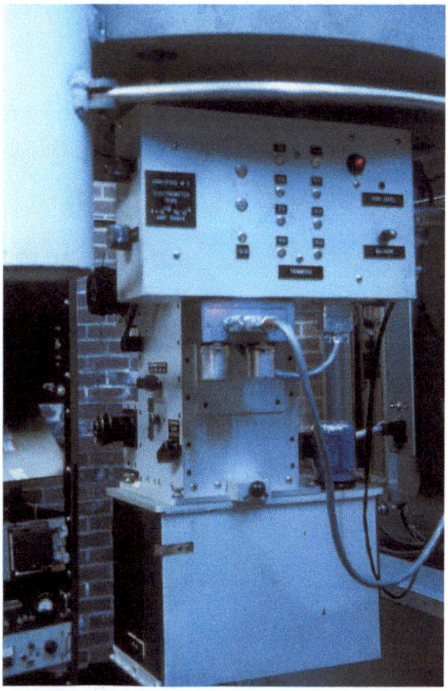

Fig. 7 The well-designed astronomical photometer in use at the Dyer Observatory, $\sim$ *July*, 1965. The rocker switch from star to sky can be seen on the lower right side of the left face, but the operation was purely manual. Note the use of desiccant containers to minimize moisture

a fixed second beam was devised by Myrabø (1979, 1980) for variable star photometry at Trömso, Norway, in the teeth of strong auroral emissions. Two apertures were located in the focal plane, at a separation of 6 arcmins; it was preceded by a rotated chopping wheel (i.e., a wheel with sectors removed) in close proximity. The two beams were imaged, alternately, on a single photomultiplier, and the chopper could be run over a frequency range of 1–20 Hz. De Biase and Sedmak (1974) discuss the principle of a multiplexed adaptive multichannel single-detector photometer that can select among a known source, an unknown source, and the backgrounds for these two sources. These principles, initially discussed by Sedmak (1973) were put to practical use in several photometers, including the twin-beam *URSULA* system (De Biase et al. 1978). Geyer and Hoffmann (1975) describe a two-photometer instrument in use at the Nasmyth focus of a 106-cm telescope at the Hoher List Observatory. This system used reflecting prism diagonals to pick off two objects, or one object and sky in the field. The system as it appeared in 1983 is demonstrated in Fig. 8 by Prof. Geyer.

Grauer and Bond (1981) describe a two-photometer system in which one photometer is located on the arm of an offset-guider and the other is placed in the main beam. The system was fabricated in duplicate for use at both Louisiana State University and the Kitt Peak National Observatory. The photometer system made use of the high-speed design of Nather (1973), but sampled the sky only briefly in about 20-min. intervals. Davidson et al. (1976) describe a two-channel photometer that employed an oscillating flat mirror that provided star and sky fluxes alternately to a single photomultiplier. The optical path was nearly identical, traversing the same diaphragm and filter.

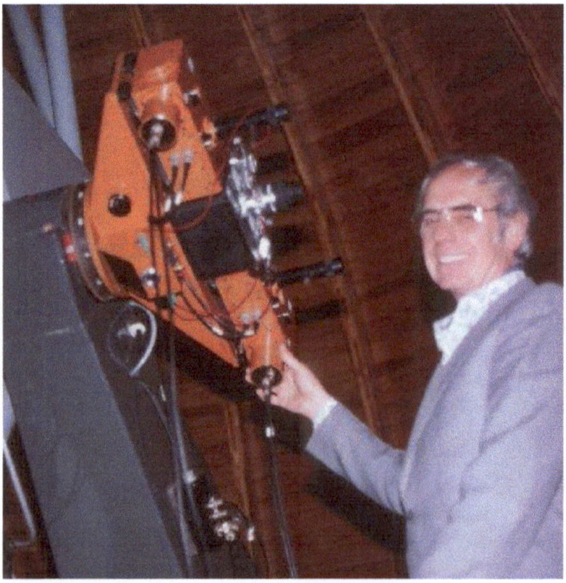

Fig. 8 The two-photometer instrument in use at the Hoher List Observatory along with its developer, Prof. Edward Geyer, ~ *August*, 1983

A high-speed photometer that employed two photomultipliers to observe both target and reference source was used at the Asiago Observatory by Bernacca et al. (1978), and another, described by Piccioni et al. (1979) was used at the Bologna Observatory.

In the early 1980s, the Rapid Alternate Detection System (RADS) was developed at the Rothney Astrophysical Observatory in the foothills of the Canadian Rockies, to overcome the cirrus-laden skies that plague western Canadian sites most of the year, but especially in the winter. The idea was to compensate for sky transparency variation while correcting for sky brightness variation, and minimizing response differences by avoiding wherever possible electronic or optical path differences. The inspiration for this device came originally from Adriaan Wesselink, who described to EFM the Walraven photometer that he had seen in action in South Africa (see above).

4.3 The Rapid Alternate Detection System

The Rapid Alternate Detection System (*RADS*) combines gated photon-counting techniques with secondary-mirror chopping techniques borrowed from infrared astronomy. The purpose in developing this system was to meet the conditions for precise differential astronomy outlined in Sect. 1.

4.3.1 RADS Hardware

The instrument as originally developed was described in Milone et al. (1982). The base photometer was manufactured by Astrometrix, Inc., of Charlottesville, Virginia, through the good offices of Prof. Laurence W. Fredrick, and was installed on the 0.4-m telescope of the Rothney Astrophysical Observatory late in 1972. Operated solely in DC-mode for about a decade, the photometer per se was described in Chia et al. (1977). Some of its design features were especially valued: the photometer head allowed access to each filter as it was rotated past an access port, and it had an exterior plug-in lamp that illumined the aperture diaphragms with fiber optics. The proposed operation of the RADS was put forward by EFM in July, 1977. Extensive discussions, first with T. Alan Clark, who with EFM co-directed the RAO, with Physics Department technician Charles H. Hanson, and then with technicians at Kitt Peak National Observatory and elsewhere, led first to an analog system, designed by Charles Hanson, whose time was contributed by Prof. Harvey Buckmaster of the Physics Department's Electron Paramagnetic Resonance (EPR) Group; a digital system was then designed and was constructed by Frederick M. Babott. Following initial trials, a pulse-counting system was installed early in 1981. In Fall, 1981, the RADS system came on line.

The flow chart of its operation is illustrated in Fig. 9, and its components can be seen in Figs. 10 and 11.

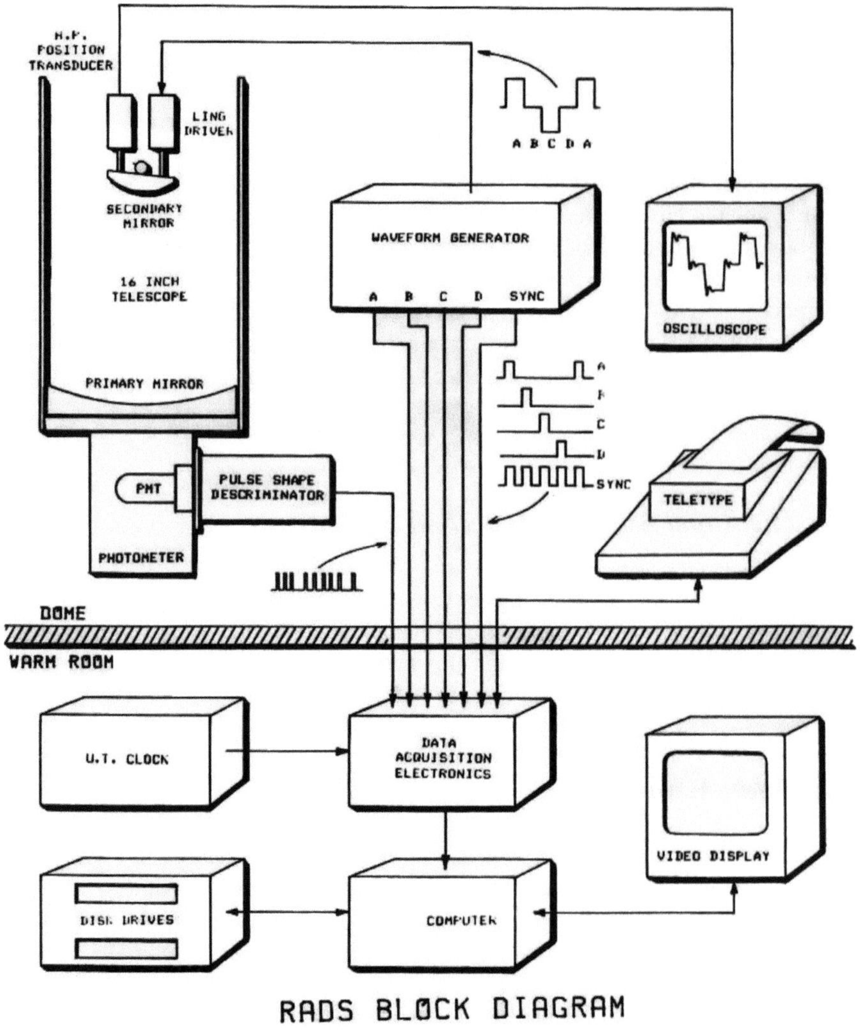

Fig. 9 A schematic and flow diagram of the operation of the Rapid alternate Detection System developed at the University of Calgary in the early 1980s by Milone et al. (1982), and adapted from Fig. 2 of that paper. See text for details

The heart of the system is basically a simple photometer of Sedmak's (1973) class A-a, wherewith the photometric targets and filters are selected sequentially, i.e., in serial mode. However, over the full integration period, the channels are accessed very nearly in parallel, and one may program the filter selection as a symmetric set. This was the usual RADS operation. At the heart of the action, a Ling Electronics electrodynamic vibrator (Model 102) drove the secondary mirror against the restoring force of two Bendix Flexural Pivots. A Hewlett-Packard linear transducer (Model 70CDT-050-B11), mounted opposite the Ling, indicated the mirror's

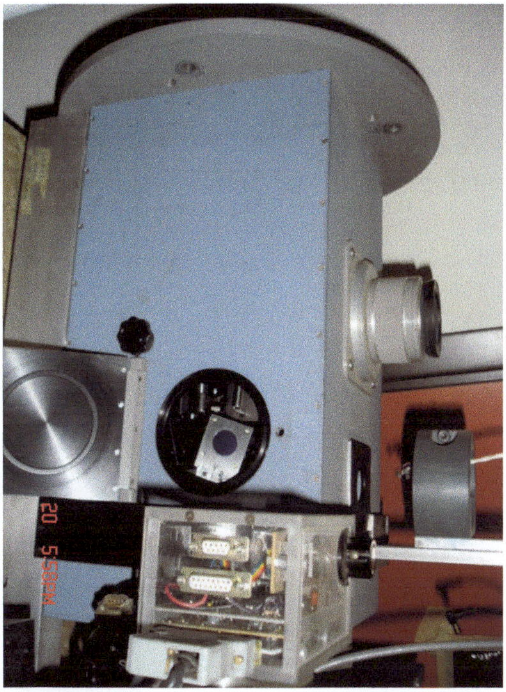

Fig. 10 Details of the RADS photometer head, showing the filter change assembly, the filter compartment through the access hatch, the field eyepiece (*above*) and CCD-imaged aperture (*below*) of the Astrometrix-built photometer

position. A function generator (the "MAC" box, described below) was programmed manually to send an amplitude voltage to the Ling vibrator for each of four mirror positions. It was used also to set the time at each location (the original range was 1–999 ms), and a "delay time" to allow the mirror shake that occurred each time the mirror was stopped, to settle down to negligible oscillation amplitude.

The entire chopping assembly could be rotated 360°, so that the chopping line could be set in any direction for suitable comparison star selection (see Sect. 2); the chop amplitude, when first implemented, achieved a maximum range of ±10 arcmin on the plane of the sky. The Ling shaker was driven by a closed loop servosystem consisting of MAC and transducer inputs, damping the Ling driver response.

The initial photometer detector, an EMI 6256 photomultiplier, was relatively insensitive (although excellent for tracking the light decline of Nova Cygni 1975 (V1500 Cygni) when it first appeared); it was subsequently replaced by an RCA C31034 (GaAs) photomultiplier, mounted in a mu-metal shield and encased in a Products-for-Research dry-ice cooled chamber, with a chamber thermometer mounted on the exterior to keep track of temperature variations.

The initial data acquisition electronics consisted of a pulse counter, four separate adders, and four time counters accurate to ±1 μs. In the early 1980s, the results were sent to an Apple IIe computer, which received also timing signals from an AstroComputer Control clock. The first power amplifier and function generator were designed and constructed by Charles H. Hanson. Later versions were constructed and maintained by Frederick M. Babott.

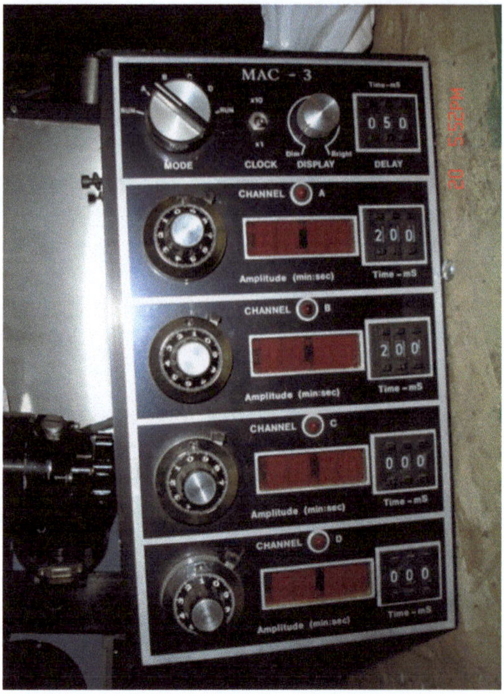

Fig. 11 The third-generation RADS "MAC box" – the multi-function generator that permits selection of duty cycle, delay time, and chop throw (amplitude of spatial variation)

In later versions, the chop amplitude was extended first to $\sim\pm 30$ arcmin, and then to $\sim\pm 45$ arcmin. Figure 10 shows the Astrometrix photometer, coldbox, and the filter and diaphragm advancing mechanism, and the CCD attached to the centering eyepiece. Figure 11 shows the Multichannel Ampltude Controller (MAC) box, the function generator that is used to select the four stations (channels) to pick up the main target (usually A), the sky near the secondary target (usually B), the secondary target (usually C), and the sky near the primary star (usually D). The sequence of filters was automated and made programmable after a few years of operation.

The output file list consisted of tables of star identifier, filter, U.T., counts in each channel, integration time in each channel, and a continuation or termination code.

4.3.2 RADS Operation and Efficiency

An observing run setup at the telescope initially required about a half-hour or more, but this decreased with experience to about 15-min under good observing conditions. The usual practice was to place the target object, usually a variable star, in Channel A, a comparison star in Channel C, and the sky in Channels B and D. Channels A through D were selected sequentially to adjust placements in a manual "static" mode and then tweaked in the "run" mode. From time to time, the dome clearance had to be checked and the dome moved; all other operations except for periodic centering checks, were carried out in the warm and lit control room. For setup, the timing controls were set so that maximum time would be spent on each channel (999 ms), to allow visual centering. The setup proceeded as follows:

- Target acquisition. The telescope was set on field of the two objects, centered half way between them. If their separation exceeded the 15 arcmin diameter of the field eyepiece, the telescope had to be moved back and forth to ascertain the position angle of the line separating them; this position angle could be computed in advance to save setup time. The required shift of direction was then noted in the aperture eyepiece.
- Chop rotation adjustment. A mode switch on the controller (MAC – see Fig. 11) would then be set to "run" to find and change the chopping angle. This angle was controlled with a bi-directional switch mounted on the telescope; the rotation was adjusted until the two lines created by the motions of the two objects overlapped, first in the field, then in the aperture eyepiece (for this exercise, the widest aperture would be selected).
- Chop amplitude adjustment. This was accomplished first by moving the mode switch from "run" to "A", and centering the main target in the aperture with the A channel 10-turn potentiometer. Then, setting the switch to "B", a sky field near the secondary object would be selected by turning the B channel potentiometer. This was followed by the selection of "C" by dialing the C potentiometer to center the secondary object in the aperture. Finally, with the mirror position switch on "D", the sky near the primary object was selected by dialing the D 10-turn potentiometer. Practical constraints on sky settings are discussed further below. The sky fields needed to be checked carefully to avoid faint stars.
- Dynamic adjustment and aperture selection. Set on "run", the centering of each channel was checked in the desired aperture diaphragm, and recentered if needed. The aperture was selected to be at least three times the apparent stellar diameter.
- Duty cycle selection. The "dwell time" at each channel would then be selected with the thumb switches for each channel. This and the delay time, discussed next, determined the chopping frequency, which we discuss below.
- Delay time selection. The thumb wheel for setting of the delay time is at the upper right of the MAC box. This time interval determined when no pulse counting was to take place. The delay time was the time allocated to mirror motion, and it included the time for the ringing of the suddenly stopped mirror to decrease to within a fraction of the selected diaphragm before the counter gate for the selected channel was opened. The required delay time could be read from the transducer trace on an oscilloscope located in the dome (see Fig. 12). The largest interval of motion was selected for the delay time. To minimize the effect of ringing, the channel settings were further constrained: The preferred setting for Channel B (sky) was about one aperture diaphragm away from Channel C, and along and within the path separating the two stars. Similarly, the preferred setting for Channel D was along the chopping line, about one diaphragm from Channel A. Because the largest excursions occurred when the secondary mirror moved from star to sky settings, the greater ringing of the secondary mirror at the sky settings resulted in minimal impact on the stellar photometry.

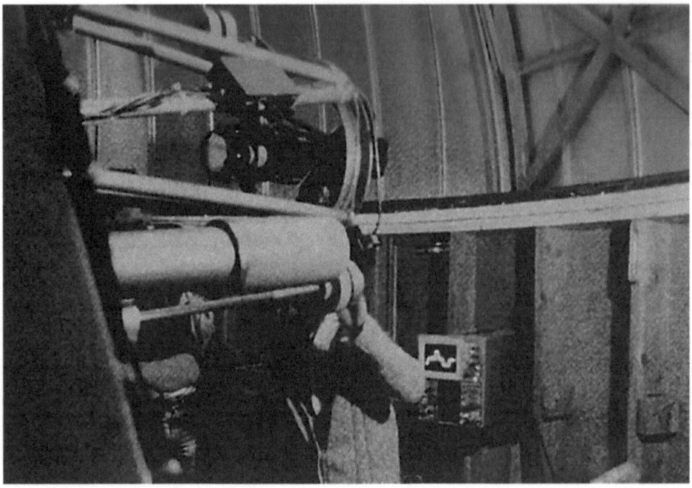

Fig. 12 The RAO 0.4-m telescope secondary mirror assembly and controller, and the oscilloscope trace showing the transducer output. Note the duty cycles of the four chopping stations. Reproduced from Milone (1993)

- Chopping frequency and integration time selections. To avoid sky variations due to, say, rapidly varying cloud or aurora, the highest possible chopping frequency should be employed; however, the higher the frequency, the lower the efficiency, in the sense that a larger fraction of any given interval could be spent collecting photons from the program star and in the other channels. This is because the parts of the cycle in which the mirror is in motion constitute overhead which must be paid in order to achieve the observational precision and accuracy desired. The integration time was set, along with the filter sequence, in the control room as interactive computer input.

Now we discuss the connection between chopping frequency and the system's efficiency. The chopping frequency of the function generator was determined by the period of a cycle, which, in turn, depended on the sum of all duty cycles and overhead settings. Thus to have a 1 Hz chopping frequency and a 1-s integration time, and if one could use a 25 ms delay time, a 225 ms dwell (counting) time for each channel could be used. In this 1 cycle example, 100 ms would be overhead and 900 ms on photon counting, 225 ms of which would be spent observing the object of interest specifically. Thus, photon counting would occupy 90% of the cycle, but only 25% of this on the target, per se. With these parameters, the percentages and system efficiency would be the same regardless of the integration time. However, if one desired to run at a higher frequency, the delay time could not be decreased much without risking signal loss, leaving only the channel duty cycles to adjust. To avoid decreasing the target star duty cycle, one's only options were: to decrease the the duty cycle of the sky channels, feasible only in a dark and nonfluctuating sky; to use only one (central) sky channel (i.e., operation in a 3-channel mode); and, reduction of the comparison star duty cycle, relative to the variable star's. The latter may be a reasonable choice only if the comparison star is significantly brighter. Usually, the duty cycles were kept the same for all channels, and more photons acquired at higher chop frequencies by selection of longer integration times. If the period of the variable star is very short, so that the integration time is phase-limited, the only recourse was to trade precision for time resolution, or move the system to a larger telescope!

For those nights when standardization could be carried out, a list of pairs of standard stars within about 40 arc-mins of each other was produced, first by Bart Milone, in the early 1980s, and later, in the mid-1980s, in a more complete form by astrophysics major Andrew Kyrgoussios; in this form it became known as the "Kyrgoussios List". Still later, pairs of Landolt stars (stars observed by Landolt in the equatorial Harvard Selected Areas) were used for this purpose (Landolt 1973, 1983). Although the RAO had a full set of *uvby* and Hβ filters, in practice these intermediate-band filters were rarely used because the wider UBVRI passbands allowed more flux to be measured per duty cycle, an important constraint for photometry on a 0.4-m telescope.

4.3.3 RADS Software Development

As one might expect, software had to be written to best serve the data as they were gathered. The initial data acquisition program was hard-wired and blasted into an *EPROM* (erasable programmable read-only memory) chip, and underwent several iterations. In the late 1980s, the interactive Data Acquisition Program for RADS (DAPR was written, in initial form by Dr. Robert H. Nelson, and later, with added capabilities and more flexibility, by undergraduate astrophysics major, Perry E. Radau. These programs yielded output lists of the data for 1–4 channels, in what was called General RADS mode; a high-speed read-out mode was available for occultation work, although the latter was rarely attempted at the RAO.

The main software reduction program, to produce systemic magnitudes, was written first for an Apple IIe micro-computer in Applesoft Basic by EFM and called CTOM (for "Counts to Magnitudes"), and later, in a slightly expanded form, and ported to a Zenith PC, CTOVMOD.

The data were then reduced with the aid of several programs. The normal procedure was to input previously determined extinction and transformation coefficients (or determine them) in the equations:

$$\begin{aligned} \Delta y &= \Delta y_0 + k'_V \cdot \Delta X + k''_V \cdot \Delta X(b-y) \\ \Delta(b-y) &= \Delta(b-y)_0 + k'_{BV} \cdot \Delta X + k''_{BV} \cdot \Delta X(b-y) \end{aligned} \tag{16}$$

and

$$\begin{aligned} \Delta V &= \Delta y_0 + \varepsilon \cdot \Delta(B-V) \\ \Delta(B-V) &= \mu \cdot \Delta(b-y)_0 \end{aligned} \tag{17}$$

where we used the familiar notation of Hardie (1962), namely, b and b_0 for observed and systemic Johnson B magnitudes, y and y_0 for observed and systemic Johnson V magnitudes, respectively, X for airmass, k' and k'' for first-order and color-term coefficients, ε and μ for transformation coefficients; in these differential formulae, the zero point terms vanish. Expressions analogous to the lower formulae in (16) and (17) were used to transform the other color indices to extra-atmosphere (i.e., systemic) and standard systems, respectively.

Except for the lower of (17), these were the equations of condition used in the least squares determinations of the unknowns. For the transformation of color terms, a variant suggested by Hardie (1962) was used instead:

$$c_s - c_0 = \mu' \cdot c_s + \zeta'_{\lambda 1,2} \tag{18}$$

where c_s and c_0 are the standard and systemic colors, respectively, $\mu' = (\mu - 1)/\mu$, and the zero points are related through $\zeta' = \zeta/\mu$, so that the coefficient to be found, μ', was a number close to zero. The quantities μ and ζ were then computed.

The program contained default values for the Sensitivity Corrections (see below) and Coincidence Corrections (required to compensate statistically for the simultaneous arrival of multiple photons within the response time constant of the photomultiplier system), but values could be read in if new determinations had been made. The k' values were computed with the Bouguer method from the sky-subtracted star data; with these new values, internal iterations could be performed.

Beginning in 1988, `CTOMVMOD` was replaced by the program `REDUCE`. Numbered versions of `REDUCE` evolved over the decade, each version with broadened capabilities and aids to facilitate its use. This work was undertaken mainly by undergraduate astrophysics majors, first James A. van Leeuwen, and later Perry E. Radau, who maintained it through the early 1990s, with some assistance from Dr. Robert H. Nelson, visiting astronomy faculty member from Prince George, B.C.. The software treated either single-channel (unchopped) or differential (RADS) data with two or more channels of output. The program applied both photon coincidence and RADS "sensitivity" corrections (from calibration of the flux through three of the channels to that in the first). The sensitivity corrections were most often determined by observing in full chopping mode the twilight sky both after and, if the system were set up from a previous night, before an evening's observation. The `REDUCE` software required input extinction coefficients unless standard star data were entered along with a list of standard star data. In the latter case, coefficients were then computed with the Hardie (1962) method.

In variable star reduction operations, the comparison star data were used to produce Bouguer extinction coefficients in each passband as well as systemic magnitudes and color indices.

Transformation coefficients were determined via least squares in a separate program entitled `TRANSFRM` which required a list of standard magntudes of the observed stars (usually taken from selected areas in Landolt (1973, 1983) tables). The Hardie method was used to obtain the transformation coefficients and zero points when absolute photometry could be performed. When the skies were photometric, through iteration a consistent set of extinction and transformation and extinction coefficients could be obtained.

The program RADSEXT was specifically written to reduce RADS observations of pairs of standard stars, in order to compute the extinction coefficients, k' and the k'' color terms in (16) and the color-coefficients of (17) and (18). Thus it was used to reduce standard star pairs, but also to provide statistical measures of the variable star and comparison star data. This was made possible because of the ability of the RADS to cancel the effect of the non-color extinction coefficient, k', on the differential photometry. The color effects that were airmass-dependent could be separated somewhat from those that were not, given a significant range of airmass of the observations. The result was such that even if the night were not photometric, in the classical sense, one could still obtain useful data on the photometry system. Examples are given by Milone and Robb (1983).

The statistical information provided by RADSEXT was:

- The mean standard error (m.s.e., or the standard deviation) of a single observation of the comparison star as computed from the differences from the mean, e_1^c, and
- That computed from the differences between chronologically successive pairs of observations of the same star, e_p^c, compared with
- The m.s.e. computed from the mean of the differences between the variable and the comparison star, e_1^d, and
- That computed from the differences between successive pairs of the differences between variable and comparison stars, e_p^d.

The next section provides an example of the usefulness of these quantities.

The Differential Variable Star Program (DVSP) produced phased light curves for further analyses and plotting.

After a light curve was extracted from the reduction programs, a program called PHIMIN, written by EFM and later ported to the PC and called EXTREME, was used to determine times or phases of minimum and sometimes maximum light. The program made use of a variation of the Kwee and Van Woerden (1956) method, in which data from opposite sides of an extremum are reflected about trial midpoints. The usage is discussed and illustrated in Milone et al. (1980). The successive numbered revisions of the PC-adapted EXTREME program were managed by undergraduate astrophysics major and later graduate student Jason R. McVean..

POLYFIT, a generalized polynomial fitting program, was used to obtain Fourier series coefficients for W UMa and pulsating star light curves.

The plotting program DCLIGHT was developed for RADS data light curves by undergraduate astrophysics major Stephen C. Griffiths, after all other software were ported to PCs. Other visualization software was made available by Douglas Phillips of the University of Calgary Academic Computer Services.

Light curves of eclipsing binary stars were analyzed with continually updated University of Calgary versions of the Wilson-Devinney program (Wilson and Devinney 1971) and Josef Kallrath's implementation of the Simplex method (see Kallrath and Linnell 1987). Since $\sim$1998, a package of light curve solvers, WDXY (where Y is a year, i.e., version), has been in use (see Kallrath and Milone 2009 for a description of this package and its evolution). The WDXY software contains its own onscreen and postscript plotting routines. Specialized software has been coded by W. J. F. Wilson to analyze light curves of pulsating stars.

4.3.4 RADS Contributions to Science and Education

The science that can be (and was) explored with this device was described by Milone and Robb (Milone and Robb 1983). Of special interest to photometrists was the possibility to measure second-order and color coefficients because of the effective elimination of first-order extinction effects. Figure 13, adapted from Fig. 1 of Milone et al. (Milone et al. 1982), illustrates the effectiveness of the system against extinction variation. The stars are the close pair HR 1764 and HR 1765. The raw V magnitude is plotted for HR 1764 (labeled "*M*") and the differential magnitude (labeled "*DY*" in the Hardie 1962 notation). Figure 14 illustrates the quality of light curve that can be obtained with RADS data. It shows the phased and processed dV light curve and $d(B-V)$ and $d(V-I)$ color indices of IR Cas obtained by RAO observer James A. van Leeuwen in 1988.

Variable star studies carried out with this instrument include that on one of the RAO's binaries-in-clusters targets in Stephen J. Schiller's (1986) doctoral program, notably DS And (Schiller and Milone 1988) in the open cluster NGC 752. Other observed systems for which analyses were completed are:

- The over-contact binaries:

  44i Boo (Robb and Milone 1982),
  TY Boo (Milone et al. 1991), and
  V781 Her (Nelson et al. 1995)

- The symbiotic binary CH Cyg (Milone et al. 1986)

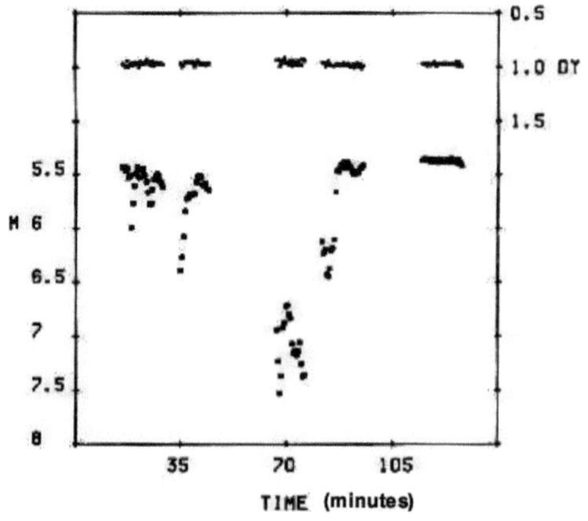

Fig. 13 Differential and single-channel light curves obtained with the RADS, The irregular variation in the latter is due to extinction caused by clouds. The raw V magnitude of HR 1764 (labeled M) and the differential raw V magnitude (labeled DY) between it and HR 1765 is plotted above. These early B stars differ by $\sim$1.0 magn. and are separated by 3.5 arcmins. in right ascension and about 2 arcmins in declination

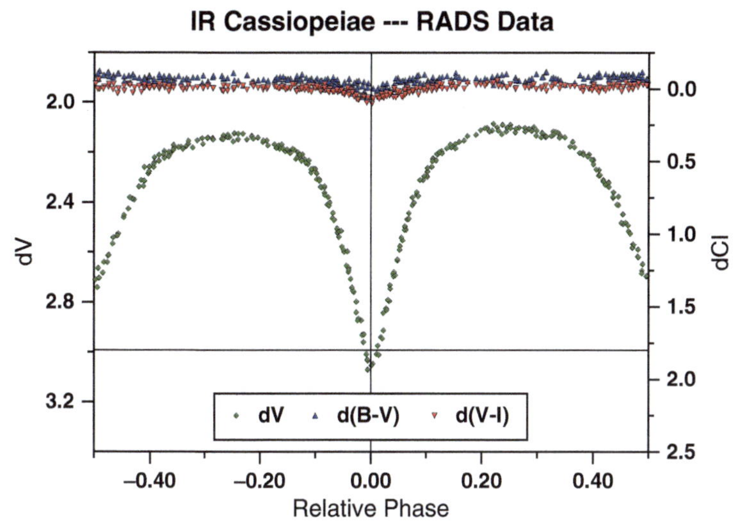

Fig. 14 The reduced and phased differential light and color curves of IR Cassiopeiae obtained in 1988 by astrophysics student and RAO observer James A. van Leeuwen with the Rapid Alternate Detection System installed on the 0.4-m telescope at the RAO

- The large-amplitude delta Scuti stars:

  EH Lib, (Wilson et al. 1993)
  DY Peg, (Wilson et al. 1998) and
  DY Her (Milone et al. 1994)

- The multi-mode delta Scuti Star V369 Sct; (Volk and Milone 2000) and
- The multi-mode RR Lyrae star AC And (Rousseau et al. (1988))

In addition to these, RADS data were obtained on the following objects; although not yet published, most of these data have been analyzed and appeared in student reports, and some of the work has been discussed at conferences.

- The eclipsing binaries QX And (H235 in NGC 752), VW Boo, CV Boo, IR Cas, VW Cep, EM Cep, RW Com, CG Cyg, V444 Cyg, RT Lac, SS Lac, UZ Lyr, and UZ Psc
- RR Lyr
- Nova Cygni 1992 and
- V376 Peg (HD 209458, extra-solar planet transit system)

In the cited paper on CH Cygni, (Milone et al. 1986) the last line of text in the right-hand column of p. 1179 was somehow omitted in the printing. We exploit this omission to make a point about science with RADS-like instruments. The omitted line reads: “May 24 data. This is evidence that the fluctuations on”. This line is in the context of a discussion of the usefulness of RADS photometry statistics for RADS observations of this interesting symbiotic binary, and in the teeth of sky background (particularly aurora) and extinction variations. What was concluded in this paper was that $e_p > e_1$ (where e_p is the m.s.e. computed from pairs of consecutive time-ordered observations and e_1 is that for a single observation computed from the mean) for both the comparison star and the variable – comparison star differences. That it was so for the comparison star demonstrated that short-term variations in sky transparency and sky background were dominant compared to longer-term extinction variation (the "airmass effect" as it is referred to in Howell's paper in this volume). The error in the differential data would be expected to be larger by a factor of $\sqrt{2}$, but in the case of CH Cygni,

> "that seen here exceeds that to be expected in both e_1 and e_p for the May 24 data. This is evidence that ... fluctuations on scales of several minutes ... are indeed present..."
>
> (Milone et al. 1986), pp. 1179–1180.

This is the type of science for which RADS-like systems excel.

The RADS instrument allowed observations to be carried out for about 30 years in an environment where portions of at most 1/3, and in bad years, less than 1/5 of the nights could be categorized as photometric. It was used by a generation of astrophysics students who learned how to carry out careful photometry and how to reduce and analyze their results. More than two dozen of these students pursued careers in science and education, and a good fraction of these have become major contributors to astronomy and astrophysics.

4.3.5 RADS Denouement

In 1997, a 1.8-m Honeycomb mirror, purchased earlier from the Mirror Lab of the University of Arizona, was successfully mounted in a support structure designed by T. A. Clark, then the newly-retired RAO co-Director and Project Manager for the ARCT mirror support project. The 1.8-m replaced a 1.5-m metal mirror that the RAO originally leased from the University of Arizona Lunar and Planetary Labs, thanks to the good offices of its director, the Rev. George V. Coyne. The mirror had been figured and polished first at the Norm Cole Workshop in Tucson and then completed at the Kitt Peak National Observatory in a project managed by the Astrophysical Research Consortium (ARC). The 1.5-m Infrared Telescope then became the 1.8-m Alexander R. Cross Telescope (ARCT), after the RAO's benefactor, who supplied the required funding to complete the mirror polishing project. The combined projects extended from 1987, when the IRT was dedicated, to 1997. From 1992–1995, the 1.8-m mirror was used in ARC's Apache Point Telescope while a 3.5-m mirror was completed, under a cooperative agreement between the ARC and the RAO. In 1999, in contemplation of a major upgrade to all the telescope systems, the RADS was moved to one of the several ports affixed to the ARCT backplane. The theory was that it could be used on much fainter objects on the larger telescope; moreover, the 0.4-m telescope was to be fitted with a CCD camera that could be accessed and operated remotely.

Despite major upgrades carried out over the following years, the RADS was never again as productive as it had been when on the 0.4-m telescope. In its new location, it was used to study several variable stars, including RT Lac and VW Cep, and to search for transits by extra-solar planets in solar-like stars. However, competition with programs requiring the infrared, spectroscopic, and CCD instruments and frequent telescope closings due to the difficulties of maintaining the technically challenging alt-alt mounting of the ARCT, diminished the number of available nights for RADS observations. After EFM stepped down as RAO director in 2004, "stare" instruments which may be maintained more easily than dynamic instruments such as the RADS, were favored, so that the RADS effectively became mothballed. F. M. Babott who had maintained the RADS instrument since the beginning retired in 2009, further limiting the liklihood of its further operation.

Despite its great promise, the RADS configuration did not catch on at other observatories, although the many references to "two-star" photometers in this article indicates that the value of differential photometry was understood. Partly the lack of adoption of RADS-like systems was due to the dynamic character of the system which required careful setup and maintenance. The characteristics of the system effectively made RADS photometry as much an art as a science. After a period of non-use, for example, the Ling driver sometimes appeared to "stick" longer than programmed at a setting, and sometimes overshoot the next position (with further use, the system "limbered up"). Moreover, settings that were set up statically, i.e., by moving the amplitude controls on the MAC box successively through the channels by hand, had to be tweaked further after the system was set to "run." Additionally, as the telescope moved across the sky, the settings needed to be checked as the Ling Driver and restoring force of the flex pivots were more efficient at some orientations than others. These and other quirks of the system meant that great vigilance was required of the observer in order to produce light curves of high accuracy and precision.

There were also intrinsic sources of error in the system, although, in principle, these could be determined and removed during analyses. For example, the photometer employed a Fabry lens to image the primary mirror on the photocathode; this is the classical way to minimize the effect of non-uniform photocathodes by having the flux illumine a fixed area. Nevertheless, the slight changes in projection of the secondary mirror at different off-axis settings required "sensitivity corrections" to the counts from three of the channels to have them calibrated to the fourth channel. This correction was typically $\sim$1% but sometimes could be as much as 5%, depending on star separation, orientation, and the characteristics of the sky in which the calibration was attempted. The use of the sky as a calibrating source also begs the question, raised earlier in this paper, about the effects of the sky background entering the detector from slightly different directions than the starlight. However, the most likely reason that the RADS did not become widely used elsewhere was the dawning of the long-awaited 2-D electronic detectors, the CCDs, in which a number (sometimes an enormous number) of comparison stars as well as a very well sampled sky, can be observed truly in parallel, if with less time discrimination.

We note, however, that photoelectric photometers that can provide rapid measurements simultaneously – or nearly simultaneously – in several passbands, are still of value, as we demonstrate in the next section.

4.4 Simultaneous Multi-Passband Photometry

Although the advantages of simultaneous multicolour photometry are evident, multichannel photometers are technically challenging. As a consequence, such photometers were not used as widely as one would expect, so that most photometric programs during the "high season" of photoelectric photometry (1950–1990) were performed with single-channel instruments.

One successful exception was the Walraven five-color photometer, discussed above; another was the Strömgren intermediate-passband *uvby* system. During the 1950s Strömgren had systematically studied and tested many intermediate- and narrow-band indices for photoelectric classification of stellar spectra. After the possibility of three-dimensional classification (temperature, gravity, metallicity) of intermediate-type stellar spectra had been demonstrated, the *uvby* system was conceived by Strömgren at Lick Observatory in 1959 and tested there the same year on the the 36-in.Crossley reflector and the 36-in. refractor. The next year Strömgren made more extensive *uvby* observations with the 20-in. reflector at Palomar. The system was soon augmented with the Hβ index developed by Crawford. The index is the ratio of light through a narrow-band filter and a broader one, both centered on the Hβ line. Large observing programs were initiated in the 1960s at Kitt Peak and a bit later also at ESO, La Silla. Early reviews of this development can be found in Strömgren (1963; 1966). During the 1970s and 1980s *uvby*-Hβ photometry became a standard tool for quantititative analysis of stellar spectra.

It should be noted that this development was not entirely based on multichannel photometers from the beginning. In the early years of *uvby*-Hβ photometry the *uvby* measurements were made with single-channel photometers and the H measurements with dual-channel instruments. The building of four-channel *uvby* photometers started in the late 1960s, notably at the Brorfelde Observatory, where the development of photoelectric instrumentation received a major impulse with Strömgren's return to Denmark in 1967. Like the five-channel Walraven photometer, the four-channel Strömgren instruments were spectro-photometers, where the spectrograph dispersion (in the *uvby* case, from a grating) provided the geometric separation of the wavelengths and the passband shapes were defined by additional filters. The higher dispersion from a grating (as compared with the prism spectrograph in Walraven's case) was needed to reduce the influence of seeing motions on the spectrum position. A complete *uvby*-Hβ measurement could now be obtained with much higher accuracy in two steps: the *uvby* data with the four-channel photometer,and the Hβ narrow/wide measurements with a separate two-channel instrument. The ultimate goal, simultaneous measurement of all *uvby*β indices with

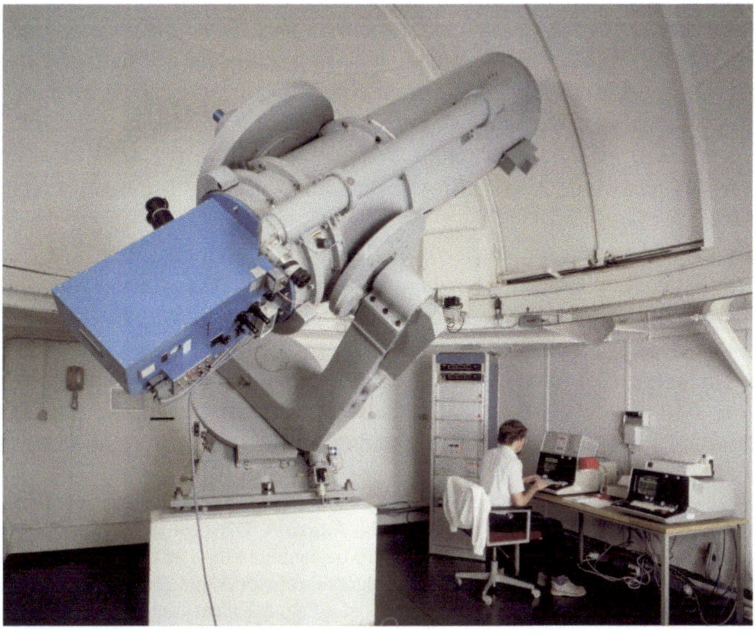

Fig. 15 The fully automated six-channel *uvby*-Hβ photometer on the 50-cm Danish telescope at ESO, La Silla (photo: courtesy ESO)

one instrument, was realized in 1987 in the fully automated six-channel Strömgren photometer on the 50-cm Danish telescope at ESO, La Silla (see Fig. 15). Further discussion of the importance of Strömgren and other system photometry are discussed by Sterken et al. and by Wing, and spectrophotometric systems are discussed also by Adelman, all in this volume.

5 Future Prospects

This paper has concentrated on the advantages of differential photometry to achieve higher astronomical precision and accuracy than would be possible without it. The technique is not flawless, and fundamental sources of inaccuracy and imprecision can still plague the photometry, especially in the presence of nonphotometric skies and ill-conceived passbands. We have focused especially on the work of Walraven and of his differential photometry legacy. We note also that Armbruster et al., in this volume, discuss a particular photoelectric system, the Pierce-Blitzstein Photometer, with which differential photometry had been carried out for more than half a century.

Experimentation can be expected to continue. In the infrared, an *IRRADS* was proposed by EFM in the 1980s, but not funded. Instead, another path to higher precision in the IR was pursued (see the Milone and Young paper in this volume).

It still seems worthwhile to implement a differential IR system such as IRRADS, however, for the highest precision attainable in the presence of partially impaired skies, if one needs to observe under such conditions in the near-infrared, or, more generally, in the mid-infrared where the ambient background is so bright that array instruments must be read out rapidly. With or without an *IRRADS*, however, the use of *IRWG* passbands is the best way to insure improved infrared accuracy and precision.

The current detectors of choice are charge-coupled devices, a separate exposition on which is given by Howell, elsewhere in this volume. The recent development of the orthogonal transfer CCD appears to be capable of producing the highest precision of this class of instrumentation.

The challenge to produce equally great accuracy remains, and to meet this challenge the skills of the observer and analyst are still paramount. We note also the importance of careful selection of passbands to minimize transformation errors.

Acknowledgements It is a pleasure for EFM to acknowledge the encouragement of Dr. A. J. Wesselink in developing the RADS, and who, with Adrian A. Disko, who had worked at Leiden in the 1950s and was employed at Yale Observatory beginning in the 1960s, provided information about the Walraven experiments. University of Calgary Professor Alan Clark provided a great deal of advice in getting the RADS program started, and additional ideas came from Professor René Racine of the Université de Montréal. Charles H. Hanson of the University of Calgary EPR Group helped realize the system. It is important to acknowledge that for nearly a quarter of a century, the management of the RADS hardware and support of its use was in the capable hands of RAO Technician Fred M. Babott of the University of Calgary Physics and Astronomy Department. Not only did he rebuild and maintain parts of instrument when needed, he often took an active role in helping students to learn its operation and the software to reduce their data. This was indeed true for all of the instruments in active use at the RAO over this interval. The Rapid Alternate Detection System was developed with the help of grants to EFM from the National Research Council of Canada, and research with it was supported later also by the Canadian Natural Sciences and Engineering Research Council. Finally we thank Andrew T. Young for critiquing an earlier version of this paper. This paper is No. 77 in the Publications of the RAO reprint series.

References

Arago, F. (1858). *Oeuvres*, **10**, 261

Baum, W. A., Hiltner, W. A., Johnson, H. L., & Sandage, A. R. (1959). *Astrophysical Journal*, **130**, 749

Bernacca, P. L., Canton, G., Stagni, R., et al. (1978). *Astronomy and Astrophysics*, **70**, 821

Bond, G. P. (1861). *Monthly Notices of the Royal Astronomical Society*, **21**, 197

Bouguer, P. (1729). *Essai d'Optique sur la Gradation de la Lumière*, Paris

Chia, T. T., Milone, E. F., & Robb, R. M. (1977) *Astrophysics and Space Science*, **48**, 3

Code, A. D., & Liller W. C. (1962). "Direct Recording of Stellar Spectra," in W. A. Hiltner (ed.) *Astronomical Techniques*, Vol. **II** of *Stars and Stellar Systems* series, (Chicago: University of Chicago Press), p. 281

Crawford, D. L. (1958). *Astrophysical Journal*, **128**, 185

Davidson, J. K., Neff, J. S., & Enemark, D. C. (1976). *Publications of the Astronomical Society of the Pacific*, **88**, 209

De Biase, G. A., & Sedmak, G. (1974). *Astronomy and Astrophysics*, **33**, 1

De Biase, G. A., Paterno, L., Pucillo, M., & Sedmak, G. (1978). *Applied Optics*, **17**, 435

Dugan, R. S. (1911). *Princeton Contributions*, **1**

Geyer, E. H., & Hoffmann, M. (1975). *Astronomy and Astrophysics*, **38**, 359

Grauer, A. D., & Bond, H. E. (1981). *Publications of the Astronomical Society of the Pacific*, **93**, 388

Hardie, R. H. (1962). "Photoelectric Reductions," in W. A. Hiltner (ed.) *Astronomical Techniques*, Vol. **II** of *Stars and Stellar Systems* series, (Chicago: University of Chicago Press), p. 178

Hermann, D. B. (2002). *J. C. F. Zöllner: Grundzüge einer allgemeinen (Photometrie des Himmels)*. Ostwalds Klassiker der exacten Naturwissenschaften, Vol. **291**. (Frankfurt am Main: Harri Deutsch)

Johnson, H. L. (1962). "Photoelectric Photometers and Amplifiers," in W. A. Hiltner (ed.) *Astronomical Techniques*, Vol. **II** of *Stars and Stellar Systems* series, (Chicago: University of Chicago Press), p. 157

Jordan, F. C. (1923). *Astronomical Journal*, **35**, 44

Kallrath, J., & Linnell, A. P. (1987). *Astrophysical Journal*, **313**, 346

Kallrath, J., & Milone, E. F. (2009). *Eclipsing Binary Stars: Modeling and Analysis* (2nd ed.). (New York: Springer)

Kinman, T. D., & Mahaffey, C. T. (1974). *Publications of the Astronomical Society of the Pacific*, **86**, 336

Kwee, K. K., & Van Woerden, H. (1956). *Bulletin of the Astronomical Institutes of the Netherlands*, **12**, 327

Landolt, A. U. (1973). *Astronomical Journal*, **78**, 959

Landolt, A. U. (1983). *Astronomical Journal*, **88**, 439

Ledoux, P., & Walraven, Th. (1956). *Handbuch der Physik*, **51**, 353

McCord, T. B. (1968). *Applied Optics*, **7**, 475

Milone, E. F. (1967). *Photoelectric Photometry of Some Selected Variables.* Ph.D. Dissertation. (New Haven: Yale University)

Milone, E. F. (1993). In K. C. Leung, & I.-S. Nha (eds.) *New Frontiers in Binary Star Research*, ASP Conference Series, Vol. **38**, p. 443

Milone, E. F., & Robb, R. M. (1983). *Publications of the Astronomical Society of the Pacific*, **95**, 666

Milone, E. F., Groisman, G., Fry, D. J. I., & Bradstreet, D. H. (1991). *Astronomical Journal*, **370**, 677

Milone, E. F., Leahy, D. A., & Fry, D. J. I. (1986). *Publications of the Astronomical Society of the Pacific*, **98**, 1179

Milone, E. F., Robb, R. M., Babott, F. M., & Hansen, C. H. (1982). *Applied Optics*, **21**, 2992

Milone, E. F., Wilson, W. J. F., Fry, D. J. I., & Schiller, S. J. (1994). *Publications of the Astronomical Society of the Pacific*, **106**, 1120

Milone, E. F., Chia, T. T., Castle, K. G., Robb, R. M., & Merrill, J. E. (1980). *Astrophysical Journal, Supplement*, **43**, 339

Myrabø, H. K. (1979). *Astrophysics and Space Science*, **60**, 367

Myrabø, H. K. (1980). *Astronomy and Astrophysics*, **84**, 297

Nather, R. E. (1973). *Vistas in Astronomy*, **15**, 91

Nelson, R. H., Milone, E. F., Van Leeuwen, J., Terrell, D., Penfold, J.E., & Kallrath, J. (1995). *Astronomical Journal*, **110**, 2400

Pel, J. W. (1991). "A Comparison of the Systematic Accuracy in Four Photometric Systems," in A. G. Davis Philip, A. R. Upgren, & K. A. Janes (eds.) *Precision Photometry: Astrophysics of the Galaxy*, (Schenectady: L. Davis Press), p. 165

Pel, J. W., & Lub, J. (2007). "The Walraven VBLUW Photometric System: 32 Years of 5-Channel Photometry," in C. Sterken (ed.) *The Future of Photometric, Spectrophotometric and Polarimetric Standardization*, ASP Conference Series, Vol. **364**, p. 63

Piccioni, A., Bartolini, C., Guarnieri, A., & Giovanelli, F. (1979). *Astronomy and Astrophysics*, **29**, 463

Pickering, E. C. (1879). *Harvard Annals*, **11**, Part 1

Plaut, L. (1940). *Bulletin of the Astronomical Institutes of the Netherlands*, **9**, 121
Robb, R. M., & Milone, E. F. (1982). *Information Bulletin on Variable stars*, No. **2182**, 1
Rousseau, C., Bourassa, J. I., & Milone, E. F. (1988). *Information Bulletin on Variable stars*, No. **3211**, 1
Russell, H. N. (1939). *Bulletin of the American Astronomical Society*, **9**, 88
Schiller, S. J. (1986). Eclipsing Binary Stars in Open Clusters: Observations and Analyses of HD 27130 and DS Andromedae, Ph.D. Thesis, U. Calgary
Schiller, S. J., & Milone, E. F. (1988). *Astronomical Journal*, **95**, 1466
Sedmak, G. (1973). *Astronomy and Astrophysics*, **25**, 41
Steinheil, K. A. (1837). *Abhandlungen der mathematische-physikalische Klasse der Bayerische Akademie der Wissenschaften*, **2**, 3
Strömgren, B. (1963). "Quantitative Classification Methods," in K. Aa. Strand (ed.) *Basic Astronomical Data, Stars and Stellar Systems* series, Vol. **III**, (Chicago: University of Chicago Press), p. 123
Strömgren, B. (1966). *Annual Reviews of Astronomy and Astrophysics*, **4**, 433
Visvanathan, N. (1972). *Publications of the Astronomical Society of the Pacific*, **84**, 248
Volk, V. A., & Milone, E. F. (2000). *Delta Scuti Star Newsletter*, **14**, 19
Walraven, T. (1949). *Bulletin of the Astronomical Institutes of the Netherlands*, **11**, 17
Walraven, T. (1952). *Bulletin of the Astronomical Institutes of the Netherlands*, **11**, 421
Walraven, T. (1952). *Monthly Notices of the Astronomical Society of South Africa*, **11**, 4
Walraven, T. (1953). "On the Use of Servomechanisms in the Photometry of Stars," in F. B. Wood (ed.) *Astronomical Photoelectric Photometry*, (Washington DC: Publications of the American Association for the Advancement of Science), p. 114
Walraven, T. (1955). *Bulletin of the Astronomical Institutes of the Netherlands*, **12**, 223
Walraven, T., & Walraven, J. (1960). *Bulletin of the Astronomical Institutes of the Netherlands*, **15**, 67
Walraven, T., Walraven, J., & Balona, L. A. (1992). *Monthly Notices of the Royal Astronomical Society*, **254**, 59
Wesselink, A. J. (1941). *Leiden Annalen*, **17**(3)
Wendell, O. C. (1909). *Harvard Annals*, **49**, Part 1
Wilson, R. E., & Devinney, E. J. (1971). *Astrophysical Journal*, **166**, 605
Wilson, W. J. F., Milone, E. F., & Fry, D. J. I. (1993). *Publications of the Astronomical Society of the Pacific*, **105**, 809
Wilson, W. J. F., Milone, E. F., Fry, D. J. I., & van Leeuwen, J. (1998). *Publications of the Astronomical Society of the Pacific*, **110**, 433
Wood, H. J., & Lockwood, G. W. (1967). *Publications of the Leander McCormick Observatory*, **15**, Part 5
Young, A. T. (1988). "Improvements to Photometry I. Better Estimation of Derivatives in Extinction and Transformation Equations," in W. J. Borucki (eds.) *Second Workshop on Improvements to Photometry*, (Moffett Field, CA: NASA Ames Research Center), NASA CP-10015
Young, A. T., Genet, R. M., Boyd, L. J., et al. (1991) *Publications of the Astronomical Society of the Pacific*, **103**, 221
Zöllner, J. K. F. (1861). *Grundzüge einer allgemeinen Photometrie des Himmels*. (Berlin: Mitscher & Röstel)

High Precision Differential Photometry with CCDs: A Brief History

Steve B. Howell

1 Humble Beginnings

CCDs were invented at AT&T Bell labs in 1969 by Boyle and Smith (1970). Their original intent was to be a bubble memory type device but accidentally it was discovered that the prototype silicon device was light sensitive and thus their fate as astronomical imagers sealed. Although it took photographic plates nearly 50 years after their discovery in 1850 to become heavily used for imaging in astronomy, CCDs arrived at telescopes in approximately 1976, only 7 years after they were discovered.

Much of the early work on CCDs at laboratories and even at observatories was related to defense applications and the space industry. This phase lasted about a decade while astronomers still regaled the virtues of photomultiplier tubes for point source photometry and photographic plates for large area coverage. Initially, only small area 2-D galaxy imaging applications seemed to find virtue in CCDs.

Some early papers, which provide interesting reading as well as historical perspective on the development of CCDs as detectors for astronomy, are those of Autio and Bafico (1975) related to the noise inserted into a CCD image by gamma radiation and Samuelsson (1975) who studied space based applications of CCDs concentrating on the effect of radiation damage to the imaging quality. Crane et al. (1979) used the Kitt Peak 4-m telescope to observe the region around the binary pulsar PSR 1913+16 and probably obtained the deepest astronomical CCD images at the time, reaching to $R = 20.9$. Finally, we want to mention the first application of "fast" time resolved spectroscopy using a CCD. Young and Furenlid (1980) and coude spectrograph obtained 13 min time resolution of the $V = 6.4$ variable star. Not as fast as one might expect for such a bright star, but keep in mind the CCD they used had a read noise of $\sim$350 electrons!

S.B. Howell (✉)
NOAO, 950 N. Cherry Ave. Tucson, AZ 85726, USA
e-mail: howell@noao.edu

E.F. Milone and C. Sterken (eds.), *Astronomical Photometry: Past, Present, and Future*, Astrophysics and Space Science Library 373, DOI 10.1007/978-1-4419-8050-2_3,

2 The Early Days of CCDs

The leaders for astronomical CCD development in the 1975–1985 time period were Kitt Peak National Observatory and the Large Space Telescope (LST) development project, mainly having its imaging work performed at the Jet Propulsion Laboratory. Both Kitt Peak and JPL had intimate relationships with CCD manufacturers and both tested new designs as well as drove even newer ideas into the next generation CCD designs. Each had fairly large groups of engineers and astronomers involved in the performance evaluation and on-sky tests of the newest devices such as the Fairchild-202 and 211 devices (Figs. 1 and 2), with rectangular pixels and fairly

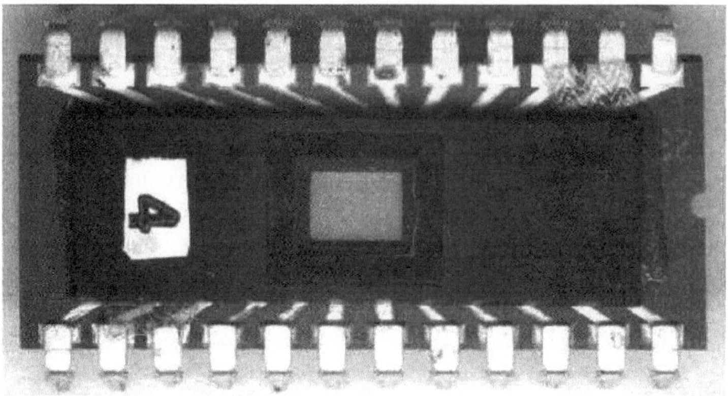

Fig. 1 Fairchild-201 CCD showing off its 100 × 100 pixel array. The CCD area is ~0.1 in. (0.3 cm) square and the entire integrated circuit is ~0.6 in. (1.5 cm) long

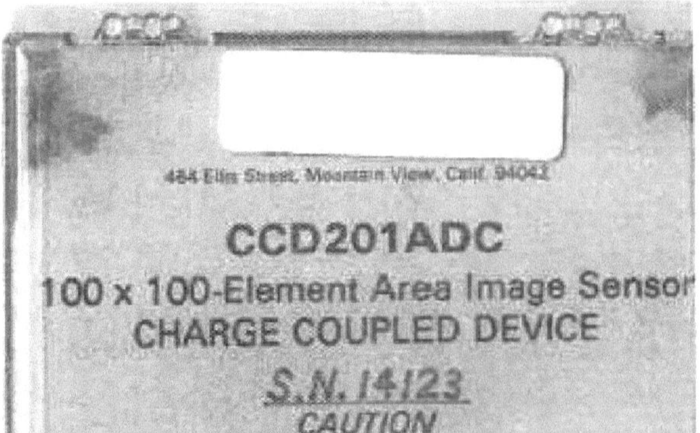

Fig. 2 Fairchild-201 CCD warning label. Note that Electro-Static Discharge (ESD) is still the leading cause of death for CCDs

high read noise, and the "large format" RCA CCDs (352 × 512) with low read noise; 20–40 electrons, a vast improvement!

At JPL, Jim Janesick led the way to astronomical use of CCDs as imagers (see e.g., Janesick et al. (1981); Janesick (2001)). Jim obtained one of the first CCD images of an astronomical object, the planet Uranus, easily showing its dark polar region (Fig. 3). This image was obtained from the Mt. Lemmon observatory just north of Tucson, AZ, high atop the Catalina mountains. The JPL "traveling" CCD camera (Fig. 4) was used to make the observation.

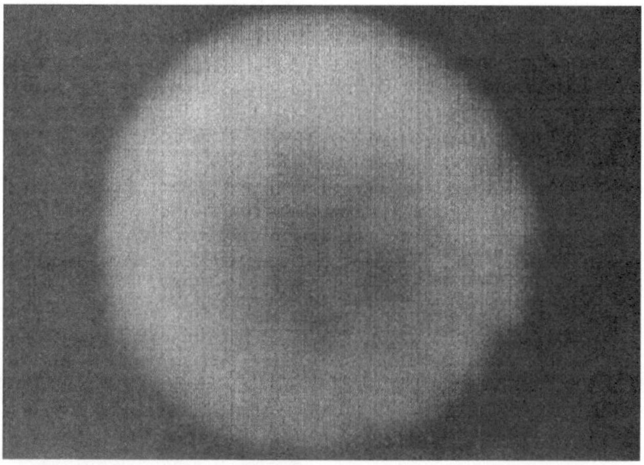

Fig. 3 First CCD image of the planet Uranus showing the dark polar region

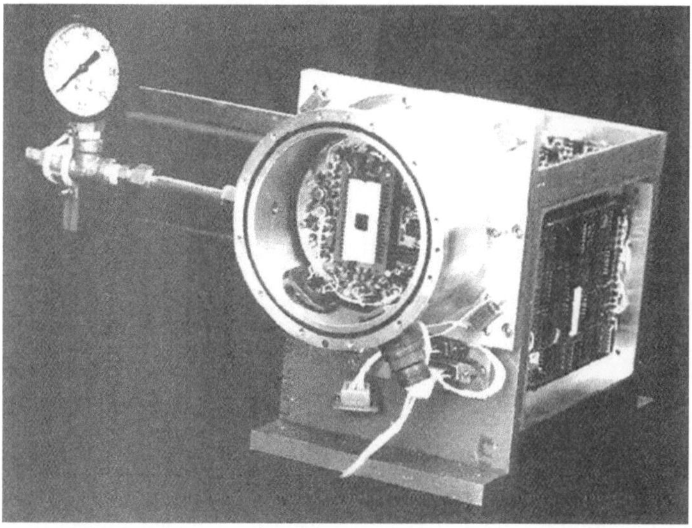

Fig. 4 JPL "Traveling" CCD camera which toured the major observatories of the time. The detector was a Texas Instruments (TI) 400 × 400 CCD

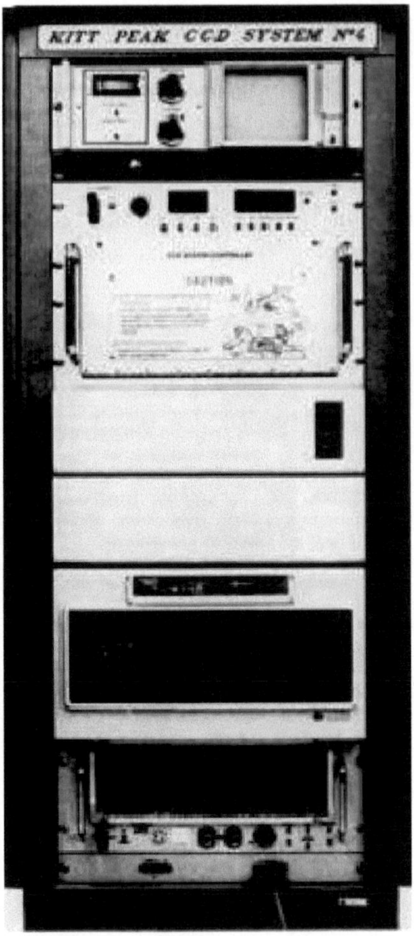

Fig. 5 Kitt Peak's 1980 vintage CCD control system. The CCD (RCA or TI) was controlled by an LSI 11/23 computer running FORTH-11. The rack contained a slow scan monitor for CCD readout monitoring, the home-built CCD controller, the LSI 11/23 CPU, a 80 Mbyte hard drive, and the electronic interface components. There even were front panel switches. Averaging together three CCD images took about 60 s and typical reduction and quick look steps took a few minutes

Kitt Peak National Observatory had their own traveling CCD system (McGuire 1983). It was intended for use by visiting astronomers at the observatory telescopes and contained more features in a stand-alone rack mounted system (Fig. 5). The system was controlled by real-time FORTH software, a language many of you reading this will be unfamiliar with, but for those select few, one which we still are nostalgic about. The rack was less portable by far than the JPL system and eventually each telescope on Kitt Peak desiring a CCD had its own rack system.

3 A Few Early Results

Since their invention in 1969/1970, CCDs were the detailed subject of many engineering and military projects and numerous early papers were published from these groups. The astronomical literature, however, was a bit slower to evolve and contained no abstracts or published papers using "CCD" during the time interval 1970–1974. However, with the introduction of the CCD cameras just discussed to the general astronomical community, published astronomical results began to exponentially rise. During 1975–1980, the word "CCD" appeared in 93 abstracts listed by ADS. In the 1 year interval (1979–1980), 34 published papers appeared in the literature. In 1981–1985 these numbers jumped to 550 and 225 respectively and during the last 5 years of that decade, ending in 1990, publications rose to 3,039 and ∼2,500.

During this early period, CCDs were still a new item in the astronomer's imaging toolbox and many held to the belief that they simply were not as good as photographic plates and photomultiplier tubes. Others, however, nearly immediately forgot these older workhorse devices and have never looked back. One of the first detailed and still highly useful papers providing a description of how to test and determine the properties of a specific CCD was that of Leach et al. (1980). Using the SAO Fairchild-202 100 × 100 pixel CCD, this paper provides a full treatment of the noise properties inherent in such an imager. The measured quantum efficiency of the Fairchild-202 was 15% in the *R* band.

McClintock et al. (1983) presented what is likely to be the first CCD time-series observations, using 15 min exposures. The marvelous TI 490 × 328 CCD was used at the McGraw–Hill observatory 1.3-m (located on the south–west ridge of Kitt Peak) to detect 0.2 magnitude variations in an 8-h period binary star system. The *R* band QE was up to ∼45% in this CCD.

Given the newness to astronomy, terminology for CCD imaging was not well established and published papers often struggled with the correct wording to use in their titles. For example, a paper on down-looking CCD satellite applications was titled "Television Astronomy" (Abramenko et al. 1984) and the original lucky imaging paper by Fort et al. (1984) performed 0.1–1.0 s exposures and was entitled "CCDs in Cinematographic Mode".

4 CCD Photometry

The beginning of CCD photometry had three major issues to deal with: the dearth of red photometric standard stars, the lack of standard photometry filters without red leaks, and any common knowledge of the methodology of how to actually measure star images from two-dimensional images.

Photographic plates, although poor in quantum efficiency compared with CCDs, were generally blue sensitive and the collections of photometric standard stars setup for use by astronomers followed suit. Hot white dwarfs with only weak spectral features and bright O and B stars were among the favorites. Standard photometric filters, such as Johnson *UBV* and Strömgren *uvby* were composed of colored glass and provided closed band-passes for the low QE blue plates but often had 1–5% transmission at 8,000Å or redder. These red leaks were not important for plates, but were problematic for CCDs. Today, standard stars used for photometry have grown into a large collection of blue and red sources, bright and faint stars, and specific stars and systems for many special purposes. The filters used today are nearly all of the interference type and red leaks are no longer an issue. Many photometric systems currently exist today, often for specific purposes; Vilnius photometry and Sloan *ugriz* are examples of astronomical photometric systems.

Whereas photographic plates were two-dimensional and indeed astronomers had been making photometric measurements with them for decades, the images obtained from CCDs had new and exciting differences. CCD data was digital from the start – that is each pixel value was unique and could be stored in computer memory for use. Thus, the astronomer had the ability to define which pixels to measure and which to ignore. subjective intensity value determinations and decisions about limits of a source in extent could be quantified. Digital photometry was born.

Additionally, those trained and familiar with PMTs now had to deal with a different kind of data product as well. PMTs provided the observer with a single number per measurement. This number represented the value one obtained through the generally large (20–30 arcsec) aperture and offsets to "blank" regions and "check" and "comparison" stars were used to sky subtract and calibrate the photometric measurements. CCD images provided the "sky" and other stars for comparison on the same image, each observed in the same manner, for the same time interval, at the same airmass, etc. New ideas and measurement techniques were needed. Today, large area CCD imaging must revisit these assumptions as many no longer apply.

Tody (1980) provided the first detailed look at how one might consistently and quantitatively measure stellar photometry using digital images. The treatise lays out a simple, yet still common straightforward method to obtain photometric measurements from 2-D CCD images. Discussions of aperture photometry and profile fitting (i.e., point spread function (PSF) fitting) were discussed as well as centroiding and noise arguments.

4.1 Absolute Photometry

CCD photometry started with doubt as to the ability to obtain precise enough observations and to match former or make new photometric measurements with good absolute certainty. Walker (1984) provided the first detailed CCD to PMT head-to-head comparison and showed that CCDs were equally good for photometric work. Accuracies of 0.01–0.02 magnitude were obtained for both the PMT observations and the comparison CCD observations. Walker used a thinned, back-side illuminated RCA CCD on the SAAO 1-m telescope for his tests.

Absolute photometric photometry with CCDs is well established today although, in fact, many of today's "photometric" observations are a far cry from the detailed robust methods and procedures truly needed to provide a fully transformed observed magnitude into one with high accuracy on a standard photometric system. Arlo Landolt, the father of photometric standard stars, provides a detailed description of this subject area in his Chapter in this volume.

4.2 *Differential CCD Photometry*

Differential or relative CCD photometry, the subject of this section and chapter, started in 1985 as a method to measure short term variations in the light output from a source with high photometric precision. Howell et al. (1985) and Howell and Jacoby (1986) provided the initial papers on the subject. Their measurements reached to faint sources ($V \sim 20$), good time sampling (10 s to a few minutes), and high photometric precision (better than 0.01 mag). All very impressive results for the time.

Astronomical sources vary due to intrinsic and extrinsic causes. Intrinsic variations are real variations of the source and the fluctuations one is desiring to measure. Extrinsic causes are variations produced by external phenomena such as clouds, seeing, atmospheric transparency changes, and systematic instrumental effects such as gain changes or other CCD electronic variations. The basic idea of differential photometry is as follows: make measurements of two or more sources in as similar a way as possible (e.g., on the same CCD image) and use differencing techniques to eliminate essentially all systematic causes of variation leaving only the signal of interest. Difference image analysis is merely a 2-D implementation of the more simple multiple source differential photometry technique.

As examples of common extrinsic variability, Figs. 6 and 7 are presented. Figure 6 shows an apparent change in brightness over time actually caused by the Earth's atmosphere (some astronomers call this the airmass effect) while Fig. 7 shows how clouds can produce spurious, random, and large apparent variations in the light curve of a source. The bottom panels in Figs. 6 and 7 show how even a simple application of differential photometry can remove these extrinsic variations and greatly "clean-up" the output light curve.

As an initial experiment (Howell and Jacoby 1986), differential photometry was attempted using trailed stars as shown in Fig. 8. Here the telescope tracking rate was slowed to allow the star of interest, U Gem, and nearby comparison stars, to trail across the CCD image while the shutter was open thereby providing a history of their variations. Extraction of the trails, in a manner similar to spectra, was performed and the comparison stars were used to normalize the source of interest into a relative light curve. Although possibly clever in design and implementation, this technique was wrought with issues and abandoned quickly.

Howell and Jacoby (1986) went on to develop the technique used today by many observers: differential time series photometry. One simply takes many CCD frames of the same object where the exposure (plus readout) time sets the light curve sampling and the uncertainty is mainly based on the S/N of the imaged stars. Early work in this technique yielded tens to hundreds of second time sampling and relative photometric precisions of tens of millimags at 18th to 22nd magnitude. The ease of this methodology, other than the large amount of image data collected for the time, provided a platform for a myriad of uses in many aspects of astronomy.

In its simplest form, differential photometry can be performed using only the source of interest, V, (a suspected variable star perhaps) and two other comparison stars usually called C and K (after the historically used comparison and check stars).

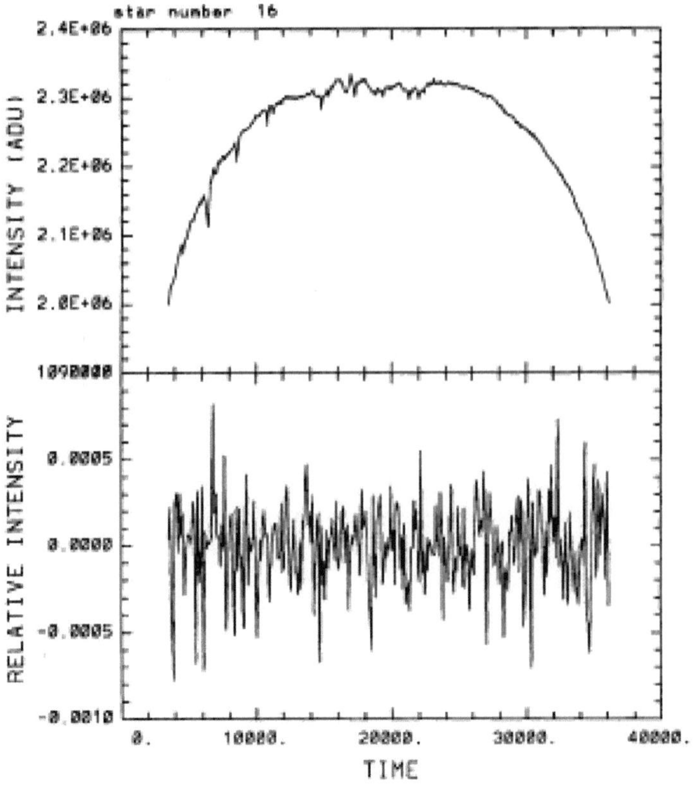

Fig. 6 Extrinsic variations caused by airmass. The top panel shows the raw flux values for a single star observed all night, starting low in the east, passing nearly overhead, and setting in the west. The apparently brightening of the star near the middle of the night is solely due to its changing airmass. Additional small dips are caused by thin cirrus clouds. The bottom panel shows a light curve for the same star, differentially corrected using other similar stars in the same field collected on the same CCD. Note how the airmass effect and the cloud dips have been removed

After often obtaining many hours worth of repeated CCD images, one measures the instrumental magnitude of each of the three stars in each CCD frame and forms instrumental light curves for V, C, and K. Forming the magnitude differences (V–C and C–K) or the flux ratios (V/C, C/K) provides simple differential photometric results (see Howell (2006)).

The formal mathematical details of how to actually formulate these differences in terms of their uncertainties per measurement and a discussion of which stars are better or worse to use as C and K is presented in Howell et al. (1988). This paper can be consulted for the methods to use to assign proper uncertainties to your differential photometry measurements.

Improvements to the simple "three star" technique were used for the open cluster M67 by Gilliland and Brown (1988). Here ensemble differential photometry was

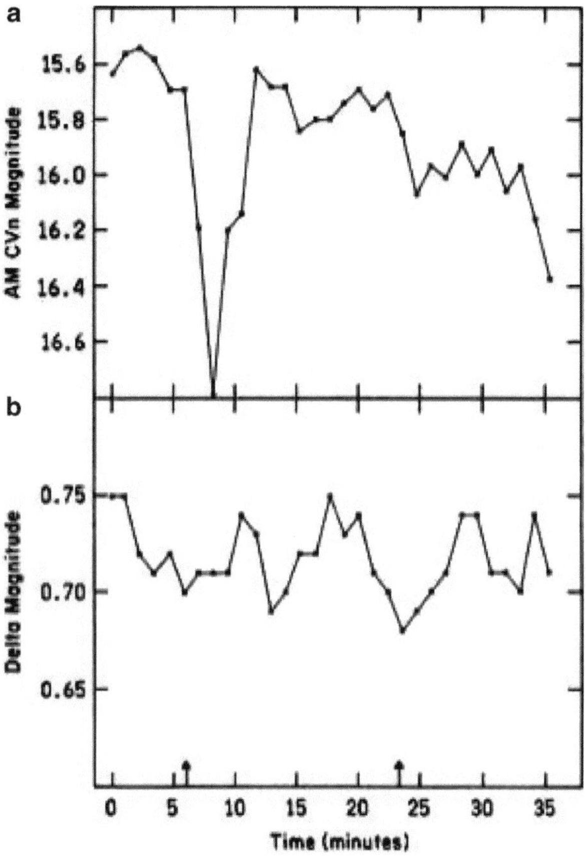

Fig. 7 Extrinsic variations caused by passing clouds. The large dip near 8 min in the top panel and the general fading near the end of the time series is caused by clouds passing in front of the telescope. After correction, in a differential sense, the bottom panel shows the flat light curve now easily revealing the 0.05 mag periodic signal for this binary star. The *arrows* mark the predicted times of primary minimum (Howell and Jacoby, 1986)

applied to many stars in the cluster using 1 min CCD images. A nice summary plot of the achieved uncertainties across the various error domains is shown in Fig. 9. M67 was a nice starting choice for a target with numerous similar brightness stars from which to build an ensemble. Inhomogeneous ensembles, including weighting schemes for the stars used, was developed by Honeycutt (1992) and the merits of local ensembles is discussed in Everett and Howell (2001) improving the photometric precision routinely obtained to at or below 1 millimagnitude.

Differential photometry is commonly produced using three photometric extraction techniques: aperture photometry, difference image analysis, and PSF fitting. Howell (2006) presents the pros and cons of each technique and we give only a brief summary herein.

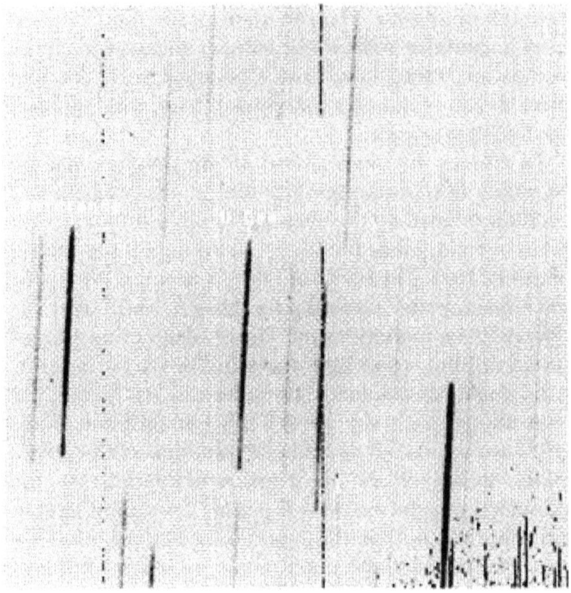

Fig. 8 Initial test of differential photometry using trailed stars. Here the variable star U Gem (*center*) is trailed along with a few other possible comparison stars. The effect of clouds on the star trails can be seen at their starting positions (*bottom* of the trails)

Aperture photometry – This technique provides photon accuracy for good photometric precision results. The use of local ensembles yields the highest resulting precision as the simple extraction method has no provision for PSF, seeing, or color changes across the possible large field of view of the CCD image. Local ensembles also greatly aid in removal of low level instrumental effects which can keep the systematic errors higher than expected. Aperture photometry is not great in crowded fields unless one uses aperture corrections which, by their nature, add systematic noise back into the results.

Difference image analysis – This technique is essentially aperture photometry with matched seeing across all CCD frames. Difference image analysis has the big advantage of producing differential light curves for all sources imaged as well as the ability to find "new" sources in the images that may show up (or go away) during the time series. The uncertainties to be aware of in this technique are those due to charge re-mapping and image transformation, plate scale changes over time and across the images, and spatial PSF differences. Generally, there is no accounting for variable CCD instrumental systematics over time or across the FOV.

PSF fitting – This technique provides the best statistical accuracy to transform instrumental magnitudes to a standard photometric system. However, it was not really developed for differential photometry and is often too cumbersome or complex for casual use.

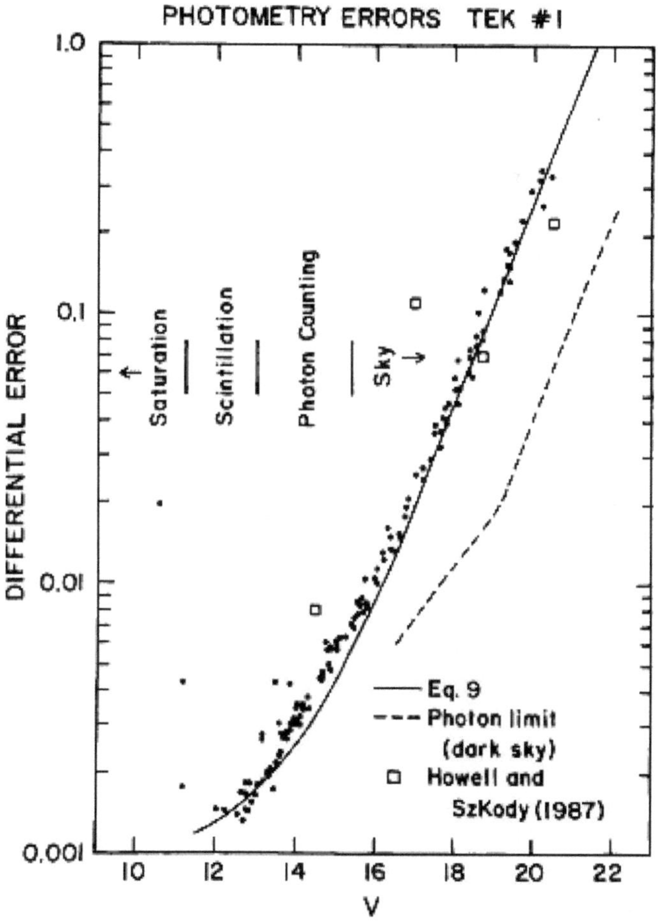

Fig. 9 Photometric error budget for the ensemble differential photometry of M67 presented in Gilliland and Brown (1988). The error budget is divided up into four domains and the *solid line* is their model error fit. (Editors' note: each type of error is omnipresent, but the domains indicate where each dominates)

The astronomical literature is filled with scientific results employing differential CCD photometry. Figure 10 shows a typical modern state-of-the-art CCD differential photometric time series dataset. Here the WIYN 3.5-m telescope and OPTIC CCD camera were used to obtain two nights of fast photometry (25 s exposures with 8 s readout time) of a ~20th magnitude short period binary system. Howell (2006) provides a more detailed review of the technique of differential photometry and gives numerous references for the reader in search of additional information.

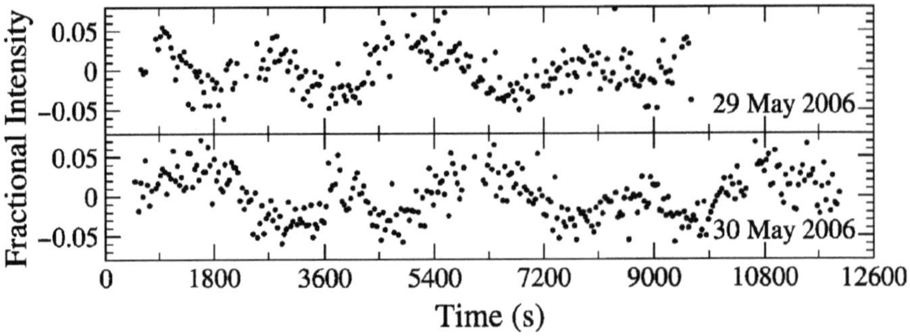

Fig. 10 Typical modern differential CCD light curve providing uncertainties near ± 0.02 in fractional intensity. The 0.05 intensity peak-to-peak amplitude of the signal is clearly seen and represents the ~ 80 min orbital period of this faint white dwarf + very late M star binary (Mukadam et al. (2007))

5 The Current State of High Precision CCD Photometry

Currently, the best ground-based photometric precisions are near 0.5 millimagnitude for stars in the $V = 8$–14 range. Much of this work has been driven by the interest in obtaining high precision exo-planet transit observations (e.g., Howell et al. 2003; Johnson et al. 2008). These works employ a new type of CCD, the orthogonal transfer CCD (OTCCD), originally developed to perform low order tip-tilt adaptive optics (AO) corrections. By the use of fast readout "guide" stars and a measured centroid feedback to the array, collected charge is shifted on the array during the science integration leading to improvement in the delivered image quality. Image improvement is similar to that obtained by a traditional mechanical AO system, 0.1–0.15 arcsec in full-width half maximum (FWHM).

However, the great advantage of OTCCDs for high photometric precision is their ability to produce square stars. Howell et al. (2003, 2005) initially explored this mechanism and obtained photometry with sub-millimagnitude precision. Johnson et al. (2008), has applied the technique to exo-planets obtaining transit light curves, sampled on minute time scales, achieving 0.0005 magnitude precision per point.

Figure 11 summaries the results for high photometric precision using differential techniques on CCD images. Over the past three decades, we have improved the precision obtained by a factor of nearly 20. Some believe now, as then, that we will not be able to improve ground-based photometric precision any further. I hope you, the reader, are among those that prove them wrong.

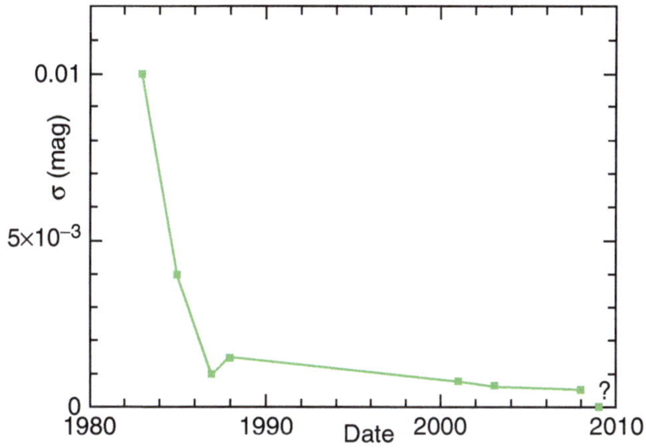

Fig. 11 Differential photometric precision improvement using CCDs over the past 30 years as evidenced by published literature values providing the best precision at the time. The 1985 value was near 0.01 magnitude while by 2008 we had reached 0.0006 magnitudes. Will we be able to gain another factor of ten any time soon?

References

Abramenko, A. N., Agapov, E. S., Anisimov, V. E., et al. (1984). Television astronomy (2nd revised and enlarged edition), (Moscow: Izdatel'stvo Nauka)

Autio, G., W., & Bafico, M. A. (1975). *Infrared Physics*, **15**(4), 249

Boyle, W., & Smith, G. (1970). *Bell Systems Technical Journal*, **48**, 587

Crane, P., Nelson, J. E., & Tyson, J. A. (1979). *Nature*, **280**, 367

Everett, M. E., & Howell, S. B. (2001). *Publications of the Astronomical Society of the Pacific*, **113**, 1428

Fort, B., Picat, J. P., Cailloux, M., et al. (1984). *Astronomy and Astrophysics*, **135**, 356

Gilliland, R. L., & Brown, T. M. (1988). *Publications of the Astronomical Society of the Pacific*, **100**, 754

Honeycutt, K. (1992). *Publications of the Astronomical Society of the Pacific*, **104**, 435

Howell, S. B., Jacoby, G. H., & Szkody, P. (1985). *Bulletin of the American Astronomical Society*, **17**, 899

Howell, S. B., & Jacoby. G. H. (1986). *Publications of the Astronomical Society of the Pacific*, **98**, 802

Howell, S. B., Warnock, A., Mitchell, K. J. (1988). *Astronomical Journal*, **95**, 247

Howell, S. B., Everett, M. E., Tonry, J., et al. (2003). *Publications of the Astronomical Society of the Pacific*, **115**, 1340

Howell, S. B., VanOutryve, C., Tonry, J., et al. (2005). *Publications of the Astronomical Society of the Pacific*, **117**, 1187

Howell, S. B. (2006). *Handbook of CCD Astronomy*. Cambridge: Cambridge University Press

Janesick, J., Hynecek, J., & Blouke, M. (1981). *Proceedings of the Society of Photo-Optical Instrumentation Engineers*. SPIE, **290**, 165

Janesick, J. (2001). *Scientific Charge-Coupled Devices*. Bellingham, WA: SPIE press

Johnson, J. A., Winn, J. N., Cabrera, N. E., & Carter, J. A. (2008). *Astrophysical Journal*, **692**, L100

Leach, R. W., Schild, R. E., Gursky, H., et al. (1980). *Publications of the Astronomical Society of the Pacific*, **92**, 233

McClintock, J. E., Petro, L. D., Remillard, R. A., & Ricker, G. R. (1983). *Astrophysical Journal*, **266**, L27

McGuire, T. E. (1983). *Publications of the Astronomical Society of the Pacific*, **95**, 919

Mukadam, A. S., Gansicke, B. T., Szkody, P., et al. (2007). *Astrophysical Journal*, **667**, 433

Samuelsson, H. (1975). *Revue scientifique et technique* A.S.E., **1**, 219

Tody, D. (1980). *Stellar Magnitudes from Digital Pictures*. Kitt Peak: National Observatory Publication

Walker, A. (1984). *Monthly Notices of the Royal Astronomical Society*, **209**, 83

Young, A., & Furenlid, I. (1980). *Space Science Rieview*, **27**, 329

The Pierce-Blitzstein Photometer

Carol W. Ambruster, Anthony B. Hull, Robert H. Koch, and Richard J. Mitchell

1 Background and Introduction

Ever since brightness estimates and measures of celestial objects began, the bugbear of ground-based determinations has been the effect of Earth's atmosphere on those determinations. Assorted hardware devices and analytical methodologies have attempted to circumvent or correct for atmospheric attenuation and selective scattering effects on measures. Typically, such concerns have been, and into the present continue to be, ignored when estimates are made.

In the 1720s Pierre Bouguer (1729) became the first person known to have attempted relative measures of the solar and lunar brightnesses by constructing a type of hardware device, the general principle of which would become popular more than 100 years later. In the early pages of his essay, he basically described a plane parallel model of Earth's attenuating atmosphere and also laid the foundation for understanding atmospheric effects on brightness. The earliest effort dedicated to brightness determinations of stars themselves was by means of the "astrometer" invented by John Herschel (1847) so that a minified image of the Moon was presented to the observer for comparison with target stars visible from the Cape of Good Hope. The difficulty of knowing the dependence of lunar brightness on phasing and the effects of librations on brightness even when phasing would be unique meant that Herschel's device would be of limited accomplishment.

The remaining years of the nineteeth century witnessed the development of numerous physical photometers, commonly of the extinction kind and primarily in Germany, England and the U.S. These telescope attachments typically used variable objective or focal-plane diaphragms, attenuation wedges, color or neutral density filters, or polarizing components (singly or in some combination) to change the brightness of an artificial source so as to bring it into agreement with that of a celestial source or to accomplish the inverse process. Development of such hardware continued through the first 40 years of the twentieth century. The summary of limitations besetting all these gadgets is very long and, for the most part, features

C.W. Ambruster (✉)
Department of Astronomy and Astrophysics, Villanova University, Villanova, PA 19085, USA
e-mail: carol.ambruster@villanova.edu

E.F. Milone and C. Sterken (eds.), *Astronomical Photometry: Past, Present, and Future*, Astrophysics and Space Science Library 373, DOI 10.1007/978-1-4419-8050-2_4,

non-matching spectral intensity distributions between the artificial source and the target star as seen by the human eye and the inevitable fact that the airpath traversed by the stellar photons can never be that of the photons from the artificial source.

Toward the end of the nineteenth century a considerable stride was made by Harvard observers under E. C. Pickering (1912) whereby the image of a target star in the northern and the northern part of the southern celestial hemispheres was brought into the same field of view as that of λ UMi. At the latitude (about 42.°5 N) of the Harvard College Observatory (HCO), the reference star varied in celestial altitude through only a limited range (from about +41.°5 to +43.°5) over a night and a year and thus its photon stream enjoyed an approximation to a constant airpath. Target stars were brought into close brightness agreement with the reference star by varying a polarizing element. For the Arequipa, Peru station, the reference star was σ Oct with an even smaller range of celestial altitude than for λ UMi. Each of the two reference stars was supported by a grid of numerous other circumpolar stars, and ultimately each has been found to be a low-amplitude light variable. Quite apart from the human eye limitations, there remained a serious problem: even though airpaths might be geometrically identical between many target stars and the reference one, these paths were usually distant from each other across the sky and the density and activity of atmospheric scatterers could be very different along the two paths.

The purely qualitative capabilities and very low quantum efficiency of the human eye as a detector could, in practice, be overcome by moving to the nineteenth and twentieth century discoveries of the photosensitive emulsion and photoconductive and photoemissive cellular devices. Over the years, the troublesome systematic effects of emulsions must have generated thousands of papers discussing possible ways to avoid or compensate for them, and the early (say, up to World War II) electrical detectors were cumbersome to use, of low sensitivity and prone to electrical instabilities which could sometimes be caused just by environmental changes.

A reader wishing to learn more ample details of these centuries of astronomical radiometry can do worse than begin with very readable 4-part summary and the modernized and more ample volume of Hearnshaw (1996), (editors' note:) as well as other papers in this volume.

The intention of this present contribution is to recite the origin, development and end of a unique hardware/software photometric system that was designed from the beginning to advance beyond the limitations of devices that had been used hitherto and to minimize atmospheric degradation of measures in an unfriendly climate. For purposes of this paper, the system will be known as the PBPHOT system, a 3-syllable acronym (with the first 2 letters individually voiced) invented originally for a piece of computer code but eventually chosen to designate the entire system.

2 The Two Prime Movers

The first two letters of the acronym memorialize Newton L. Pierce (1905–1950) and William Blitzstein (1920–1999) (Fig. 1). None of us ever met Pierce, but the eldest author was introduced to Blitzstein in 1953 and the rest in the 1960s and

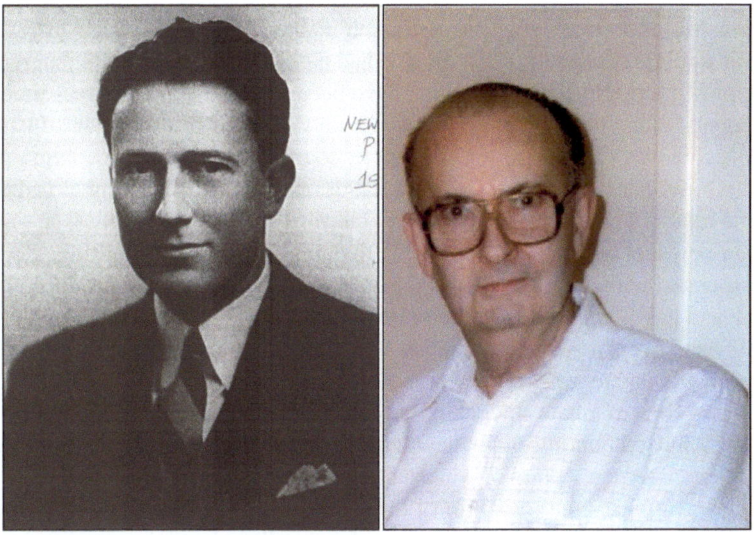

Fig. 1 *Left:* Newton L. Pierce in his mid-40s. *Right:* William Blitzstein around the age of 60

1970s. A small ingredient of what follows is oral history from him and personal experiences but the greatest fraction has been assembled from Blitzstein (1953, 1958, 1988) and from a large amount of paper and computer disk documentation, formerly observatory files now in the hands of RHK and RJM.

The main observing program at the Princeton University Observatory (hereafter PUO) during the early part of the twentieth century was that of Raymond S. Dugan (1878–1940) who used a polarizing photometer with the eye as a detector in order to accumulate light curves of many eclipsing variables. The Princeton instrument was a copy of the polarizing photometer in use for many years at the Harvard College Observatory. A Princeton PhD and then a tenured faculty member there, Pierce continued this program after Dugan's death. Their well-regarded data and analyses appeared in several *Contributions from the PUO* culminating in the doubly-posthumous publishing by Wood (1951) of still more results by the two older men. Although not so productive as Dugan, Pierce had ample experience to regret that the polarizing photometer was physically a very demanding instrument if credible observations were to be attained. In addition to the eye fatigue, measures had to be recorded by hand so that productivity was limited and, unless the night were one of uniform transparency, random errors were introduced by atmospheric fluctuations and, most likely, these could not be quantified so as to remove their effects. There was motivation enough to move to simultaneous observing of program and reference stars and to install modern detectors and recorders so as to mitigate these problems if the PUO variable star program were to continue. Pierce envisioned accomplishing these changes.

As a commuter student, Blitzstein accumulated his academic degrees from the University of Pennsylvania (hereafter UP) but was not interested in the visual and photographic observing programs sustained at the suburban Flower Observatory. While an undergraduate Physics major, he had passed through the junior/senior level course in Modern Physics which must have been updated almost in real time with the students as captive human subjects. As an Assistant Instructor and Instructor in Physics he was exposed to the concept of counting particle and radiation pulses by vacuum tube instrumentation in national defense projects. Having already mastered the required course work in both Geometrical and Physical Optics, he saw that this then-new technology could be turned to determination of stellar fluxes, and the labs associated with the Electricity and Magnetism semesters had introduced him to vacuum and gas-filled diode detectors. From his teens he had cultivated a laboratory mentality and even then was regarded as an expert in optical figuring and silver coating among the amateur astronomers in the Philadelphia metropolitan area. In 1945 while a part-time graduate student in Astronomy, Blitzstein had articulated the rudiments of a pulse-counting radiometer. (In the following year, Kron (1946) gave a kind of *gedanken* description of what such an instrument could accomplish for a selected observational situation). From 1941 to 1947 Blitzstein worked as a consulting physicist for a small company that designed and built industrial control and testing instruments and so expanded his instrumental capability while from 1947 into 1950 he was a part-time grad Student Assistant in the UP Astronomy Department with duties to develop a pulse-counting astronomical photometer.

It is unclear how information passed between Philadelphia and Princeton, but Pierce and John E. Merrill (1902–1991) visited the Flower Observatory in 1947 for Blitzstein to give them a demo of the circuitry that he had breadboarded by then. The two men found congenial interests and complementary aptitudes and went to work. Although the cellar of the Flower Observatory could have been used, they chose the space that Pierce had available to him probably because more testing apparatus and machining tools were available at Princeton. When from 1947 to 1950 Blitzstein was a part-time Research Engineer at the Franklin Institute Research Laboratories, his assignments were in aircraft fire control and analog computing and simulation. Through these three years, he commuted by rail one evening a week to Princeton Junction where Pierce picked him for several hours work. This habit ended with Pierce's death.

3 Enablers, Both Forthcoming and Inadvertent

No significant governmental financial support existed in the 1940s because the National Science Foundation (NSF) was not established until 1950 but money and support in kind had to be found to realize Pierce and Blitzstein's aims. It so happened that Blitzstein was an accomplished scrounger. As mentioned above, he had held industrial and research positions through the 1940s and for a few years thereafter was a full-time Research Engineer at the Franklin Institute Research Laboratories and an Electronic Scientist at The Frankford Arsenal of the U.S. Army. Until 1977 he was also a part-time consultant to a considerable variety of commercial enterprises and federal entities as UP permitted moonlighting at the rate of one day per week if it contributed to one's professional advancement. It would happen that Blitzstein's eye would fall on this or that piece of hardware at a lab or business and he would offer to examine and test it on his own time at the UP observatories. Large items and small ones appeared in this manner and sometimes made their way back to their original homes. Presumably others were written off as surplus.

One cannot live by sticky fingers alone; real money was needed from the beginning and this was usually to be had although commonly in sums that might have been larger.

The first person to give support for the eventual PBPHOT system was the UP Astronomy Chairman, Charles P. Olivier (1884–1975). Olivier had the personal manner of a crusty southern aristocrat but he was a very fine scientist of more than a little breadth. Although his deepest specialty was the meteor phenomenon, he was also an experienced variable star estimator, primarily of cool giants and supergiants. He had no regard for the astrometric program that had been the Flower emphasis up to his own appointment and essentially junked that (still valuable and useful) hardware so it might be considered somewhat surprising that he bought a photographic wedge photometer for grad students and volunteer observers to move to measurements beyond his own capability. His putting Blitzstein on the payroll for the development of a pulse-counting device has to be seen in the same light: although he couldn't do some things and perhaps didn't understand them fully, he was willing to be sold on a new enterprise if it had scientific promise.

There were times when loose change simply didn't exist in the departmental budget after World War II. At a few of those times, an unlikely source was tapped. I. M. Levitt (1908–2004) was an astronomy graduate student and Blitzstein's contemporary and friend. In a technical sense, Levitt was not Blitzstein's equal but he was a good pair of hands in the lab and a good observer and so assisted Blitzstein in tests and experiments. He acknowledged being grateful to Blitzstein for teaching him enough subject matter to get through their Celestial Mechanics course. In return, Levitt coached Blitzstein along so that he could pass the required German exam. Furthermore, Levitt had some resources such as his father-in-law and successfully appealed to that man when money was short. Levitt's marriage survived, and he developed a very successful career in the public education program sited at the Franklin Institute and Fels Planetarium.

Pierce and Blitzstein had observing experience only with their institutions' long-focus visual refractors, the Princeton one being the larger and newer instrument. By the time they began their collaboration, they knew that other telescopes could eventually be available to them. These had been the property of G. W. Cook (1867–1940), a very prominent businessman who, beginning in 1935, had built a well furnished personal observatory on his Roslyn Hill estate in a Philadelphia suburb. He had installed a 15-in. stationary horizontal refractor – a siderostat – and a 28.5-in. Newtonian/Cassegrain reflector as well as several other smaller instruments in his buildings. Not being able to keep all this hardware productive himself, Cook encouraged capable amateurs and professional astronomers to ask for telescope time for their own purposes and appointed Orren C. Mohler (1908–1985) as his head scientist. This excellent choice resulted in good science produced and published but everyone knew that Cook was not so healthy as he might be and East-Coast astronomers wondered hopefully what would happen to the equipment and station when he passed away. This was answered in his will which deeded all the hardware to UP provided it was removed from the Cook property within a reasonable length of time after his death. Olivier was privy to these details as early as 1937. It must be considered a coincidence not worthy of note that UP gave Cook a degree *honoris causa* a couple of years before his death, the first such degree awarded to a person without at least an earned BA or BS. World War II and the nearly closed university research program meant that the removal timetable could not be met. Most graciously, Cook's widow, Lavinia, permitted the station to remain as it had been, keeping up the grounds and even permitting continuing research use until UP could come up with the people, time and funds for moving the equipment and razing the structures.

In addition to the individual tragedy of Pierce's death at an early age, practical concerns had to be faced. In the first place, there was no one at Princeton interested in sustaining Pierce's program and the recruiting process for a new faculty member was unlikely to find a comparable individual. Secondly, Blitzstein's appointment as a consultant to the PUO had to terminate. Lastly, Princeton had put money into the instrumental development and retained title to it. Two individuals dealt with these matters in a congenial and collegial way. In August, 1950 and just 8 days after Pierce's death, Lyman Spitzer, Jr. (1914–1997) wrote to Blitzstein from the West Coast asking to be brought up to date on the status of what had been constructed, how much more needed to be done, whether components could be cannibalized for other purposes, and whether the PUO might retrieve any expenses. Blitzstein's response does not survive if he made any, but he clearly turned the inquiry over to F. B. Wood (1915–1997) who had been appointed to the UP faculty in 1950 as Executive Director of the observatories before Olivier's retirement. He would be Olivier's replacement as Department Chairman as well. Wood was interested in eclipsing variables and his Princeton dissertation (Wood 1946) showed that he could create light curves visually, photographically and photoelectrically and could model them well. After U. S. Navy service during the war, he had been appointed at the University of Arizona where he sustained his original interests. Now he determined to acquire the products of the 4 years of the Pierce/Blitzstein collaboration and bring it to completion and telescopic productivity. To this end, he quickly got in touch with Spitzer to ask what Princeton would accept to surrender title to the developed hardware and associated test equipment. Spitzer's answer spoke of $8,500 as a fair value and in mid-September Wood approached Glenn R. Morrow (1895–1973), Dean of the School of Arts and Sciences and a most eminent classicist and philosopher, to see how this sum might be found even though the departmental budget had been fixed 5 months earlier. Wood's letter is direct and forceful, emphasizing the modernity and expected longevity of the instrument when it would be completed, as well as its suitability to an eastern U.S. climate. Before October 4 Morrow had found the sum and even came up with $10,900. Spitzer then promised physical transfer of the apparatus by January, 1951 and this happened as planned.

It was then up to Olivier and, more pragmatically, Wood to lay hands on the Cook furniture and install the beginnings of the PBPHOT system at a new station so that observational work could begin. The stops and starts of the creation of the Flower and Cook Observatory are not part of the present story but the place was functioning by 1955.

4 The Pierce/Blitzstein/Levitt Years

While Pierce sustained his teaching duties and eclipsing binary research and published the 2nd edition of *A Finding List for Observers of Eclipsing Variables* (Pierce (1947)), Blitzstein continued his full- and part-time research and consulting jobs. Levitt and he presented their first designs and test results at the 77th AAS meeting in 1947 where A. E. Whitford encouraged them to push along with development. This they did and Blitzstein offered a progress report at the next AAS meeting. Some stimulus to move faster was provided by Yates's (1948) summary of a single-channel, pulse-counting photometer that had already been put into observational service at Cambridge, U.K. When attached to an old telescope diaphragmed down to 7 in., it was easily able to make precise measures of Uranus.

As some evidence of their progress, there is the UP publication by Levitt and Blitzstein (1947) which attempts to explain the theory and practical aims of photoelectricity and pulse counting to alumni who have no technical background. In the posed photo (Fig. 2), Levitt is pretending to acquire a star with the Flower 18-in. refractor while Blitzstein is seeming to use a stopwatch to ensure uniform counting intervals. In front of him there is a counter displaying numbers of pulses, which numbers he would write on the pad. There is a power supply on the floor and a vacuum tube amplifier on the table. It is all there in a rudimentary form and they appear not to be in Cambridge's league. The single-channel photometer that Levitt is touching was made in the machine shop of the Franklin Institute. This same article also takes note of some people at the Franklin Institute and at RCA who gave more than a little technical help and implies that Pierce had been in touch with the UP people from some earlier time. In the same year, other magazines such as *Radio Age* and *The Institute News* and the national daily *The Philadelphia Bulletin* took note of the work of the two graduate students. Levitt received his PhD this year (1947) for wedge photometry of the eclipsing variable ZZ Cas.

Subsequently at Princeton, construction, acquisition, and testing of more advanced and impersonal modular components for the entire system were the main emphases. Double binary pulse counters as well as a seasoning chamber to keep multiplier photocells under voltage in the dark were constructed by the two men. With Princeton money, they bought a temperature-stabilized tuning fork whose frequency could be re-scaled to 10^{-5} days, which was intended to be the unit of time for astronomical counting. Later they changed their minds in favor of more flexible time scaling. Other purchases included electro-mechanical printers for recording

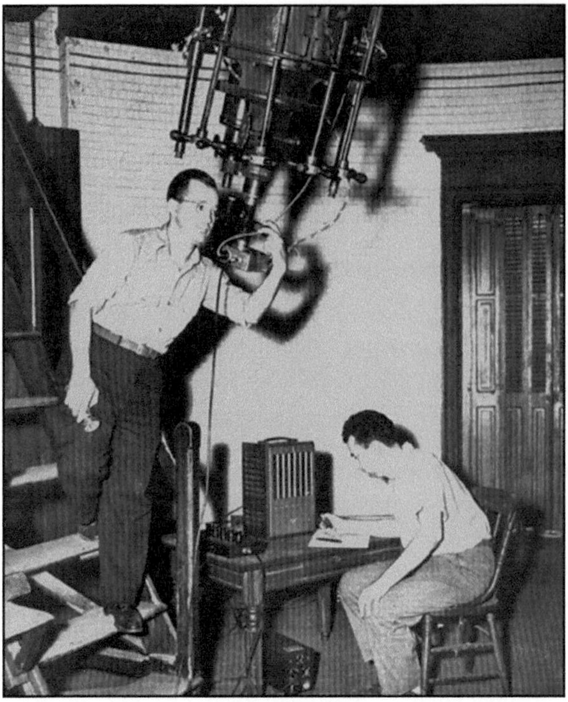

Fig. 2 Levitt at the focal plane of the Flower Observatory 18-in. refractor with Blitzstein at the table. Primitive versions of all the essential hardware components of the photometer are arrayed around him. The photo is a posed one taken in the daytime and was the cover image for the issue of *The Pennsylvania Gazette* which contains the Blitzstein (1950) contribution. *The Pennsylvania Gazette* is published by the Alumni Society of the University of Pennsylvania and the image is copied here with permission from the Society, which permission is gratefully acknowledged

time, counts and target IDs on paper tape and numerous photocells, mostly RCA 1P21s, to replace the outmoded 931A tubes that had been the earliest detectors. These hardware items were assembled into a packaged system, tested for its limitations, changed to improve it, and then tested again in a continuing cycle to try for a system that would be fail-proof in the dark.

What was not accomplished at Princeton was to move beyond a test bed for a single-channel device. Although the ideal of a dual-channel photometer remained the aim, there was not time enough to give thought to realizing the mechanical mounting for two channels onto a telescope. As it happened, this would probably have been work and time wasted, for they remained unsure what would become the preferred telescope.

In addition to working full-time and commuting to Princeton once a week, Blitzstein took time to make the observations for his dissertation. He benefited from Lavinia Cook's indulgence by building an attachment for his and Levitt's single-channel pulse counter so as to attach it to Cook's stationary, horizontal

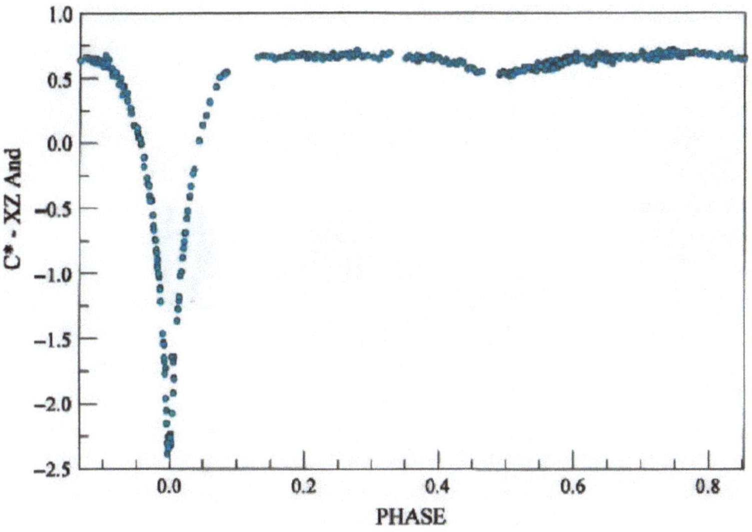

Fig. 3 The first pulse-counting light curve of a variable star. This is copied not from its original publication in *The Astronomical Journal* but from Blitzstein's notes

visual refractor. He never had a driver's license during his entire life so public transportation and a half-mile walk brought him to the Roslyn House facility in 1948 and 1949. Because the newer hardware had not yet been acquired from Princeton, he worked very much as is indicated in the photo with Levitt. The result was the first pulse-counted (unfiltered) light curve ever obtained and it was of the eclipsing binary XZ And, which drops to about $V = +13.0$ at minimum light (Fig. 3). Published quite promptly as Blitzstein (1950, 1954), it brought the Princeton-UP collaboration to more than parity with Yates's group.

5 Until 1955

In 1950 Wood instantly had Blitzstein appointed a part-time Research Associate to the observatory while he continued his day job. Blitzstein's task of re-assembling the components from Princeton was easy but further quick progress was limited to designing and building a better mount that would be necessary for the PBPHOT to be attached to a telescope. That single task was not completed until 1954. In the meantime, a provisional observing program was set in train with the Princeton circuitry being coupled to the rebuilt single-channel photometer. A photo accompanying another local publication (Wood 1952) shows the Director hemmed in by an ensemble of chassis with nothing being rack-mounted, and he is looking at the re-built old photometer fed by Cook's horizontal refractor.

The Blitzstein (1953) contribution is just one of 8 that Wood had assembled for a December, 1951 symposium sponsored by Section D of the AAAS. The entire short volume is worth perusing for its thinking at the time as well as for Blitzstein's summary of how far he and Pierce had come and what he believed to be in the near future for himself. It is also a good example of the way in which Wood was trying to increase visibility locally and nationally for astronomical science. The volume's title is *Astronomical Photoelectric Photometry*.

In 1954 a pair of academic appointments brought Blitzstein onto the full-time UP academic staff: He became Assistant Professor of Astronomy and Astroelectronics in the College of Arts and Sciences and, with the assistance of John G. Brainerd (1905–1988), Dean of the Moore School of Electrical Engineering, also in the School of Electrical Engineering. From then on he climbed the academic ladder at a reasonable rate. Without outside distractions, he continued to assemble and develop the 2-channel apparatus and, although he had a dedicated machinist, he did not yet have an electronics technician. Even when the new Flower and Cook Observatory (the FCO) was opened, he still had no such assistant.

The picture of Wood toying with the photometer implies a decision that had already been made. When the new FCO would be functioning, the then-existing photometric system and its soon-to-be replacement, the PBPHOT, would not be mounted on the Cook reflector. Rather, more than a factor of $4\times$ in light-gathering power would be surrendered by mounting the system on the smaller 15-in. stationary, horizontal visual refractor. There were a few reasons for this seemingly illogical decision.

1. The reflector was not a sturdy fabrication despite it having been reinforced by Mohler years earlier whereas the focal plane of the siderostat was defined by a large, wall-mounted steel casting complete with setting circle displays and the capability of mounting hundreds of pounds in attachments.
2. The focal planes scales of the two telescopes ($20''\mathrm{mm}^{-1}$ and $39''\mathrm{mm}^{-1}$ for the reflector and refractor, respectively) favored the reflector but there was another detail to consider. The small central hole in the objective at the reflector's Cassegrain focus made Wood and Blitzstein fear that they would not be able to find comparison stars close enough to many program stars so that two objects could be observed simultaneously. For the siderostat, on the other hand, the accessible focal plane diameter was a full $1^{\circ}\!.6$ and it would surely be possible to find at least one comparison star over that diameter.
3. There was a functioning multi-filter photometer already adapted to the reflector for which they had to spend no money and its busiest user didn't want to move his instrument to the smaller telescope.

The siderostat design permitted no finder to be generally collimated with it so acquisition of the intended star field meant that the observer had to use that telescope alone. Further, the Brashear objective was necessarily a visual doublet because the alternative photographic doublet gave such out-of-focus images that the observer could not easily recognize star fields. So there was the inevitable second limitation of secondary spectrum in the images presented to the photometer and this imposed larger-than- wished-for focal plane diaphragms in the PBPHOT.

A third consequence always had to be reckoned with as well. The optical components were not in the building with the observer but in a conical-roof shack north of the observing room. The optical beam was contained inside a steel tube connecting the buildings. If the observer wanted to check that the flat really had an unobstructed view of the sky, he had to go outside to the telescope shack and stick his head under the roof shutter to look at the sky. The roof and shutter had to be rotated by hand.

For any Pollyanna there could be mitigations of these concerns. The observer was inside a room whose temperature and relative humidity could, in principle, be controlled, all the equipment was within easy visibility and hand reach and it too was not subjected to climatic extremes, and, if something went wrong, replacement components were stocked at hand. Of course, the observer couldn't see the sky as he observed and one had to be confident he understood what he was doing if repairs were attempted.

The telescope log for the 1951–1955 interval when the prototype instrument functioned at the Roslyn House station can no longer be found. Some information exists in the memory of the oldest co-author who did his first photometry with the prototype single-channel photometer at that time and achieved some published timings of eclipse minima. Wood and Blitzstein exercised the hardware and software on numerous stars so as to test the system as it existed then but a systematic observing program cannot be said to have existed. Workers were looking forward to Blitzstein installing his full design at the new FCO.

The final program accomplished with the pulse-counting system at the Roslyn House Observatory was the monitoring of the 1955–1956 eclipse of ζ Aur by Wood and Blitzstein (1957). It is worthwhile to give a feeling for how they had to work. For an unknown reason, they decided to observe that eclipse in sequential mode rather than observing the program and comparison stars simultaneously. The observer set the first of 4 color filters across the beams and, had the filters been wide-band rather than interference, an additional neutral density filter would have had to be used for ζ Aur itself because that star was so bright that the pulses could not be time-resolved by the counter unless the beam was so attenuated. Voltage discriminator levels had already been set individually so as to avoid counting the low-voltage pulses due to thermionic emission and the counting interval chosen to give an acceptable S/N ratio. A WWV signal was then entered manually into the record so that a time base could be recorded automatically thereafter. A switch was tapped to initiate the counting interval, and the star's identifiers, filter codes, and pulse counts were entered manually into an adding machine device when the counting interval had run its course. The next measure, possibly with the same filter or possibly with another one, was then ready to begin. The two paper tape records were transcribed the following day (sometimes) and the measures reduced with a Monroe or Friden electromechanical desk calculator.

6 The Early FCO Photometrically

In order to open and furnish the FCO it was necessary to sell the Flower property and design and erect the new structures. Insofar as these are known, the false starts, compromises and decisions about these matters are related on the web[1]. The installation of the Cook siderostat was particularly challenging and demanded endless day and night hours of Wood, Blitzstein, Merrill (now on the staff) and William M. Protheroe (1925–), a junior UP faculty member. It was a fussy task and they didn't get it quite right by failing to adjust the optical centers of the flat and objective in a horizontal line and by locating the optical axis about 1 in. too far to the west in the telescope room.

The 15-in. objective was still fed by Cook's 30-in. diameter, altazimuth-mounted aluminized flat. As the flat rotated to follow a star field, it could become so steeply projected as seen from the objective that constant star flux would not be captured by the objective and spurious variability would be introduced into the measures. This problem was most troublesome at the north extreme of declination. At about $\delta = +66^\circ$ essentially no flux was presented to the objective by the flat because it was lying nearly horizontal and more northerly fields could not be seen at all. As if this limitation were not sufficient to test the dedication of the observer, there were a few more compounding annoyances. At declinations approaching $+66^\circ$, the radiation reflected from the flat became sensibly polarized and the polarization level varied with the angle of incidence on the flat. Wood also wished to remove from an observer's mind the temptation of observing very low in the sky, principally to the east where Philadelphia lit up the sky so he had the walls built higher than was needed to house the telescope, further limiting sky visibility.

The installation was dedicated at a 2-day symposium in June, 1956 and the Blitzstein (1958) contribution appears therein. The symposium topic, *The Present and Future of the Telescope of Moderate Size*, expresses the view that photoelectric detection will expand greatly the capability of modest telescopes although not all the contributions turn out to be concerned about that matter. There is also some play given to the concept of automation in real-time observing. Beatrice, Pierce's widow, was an honored guest at the event and was photographed with Blitzstein at the optical head of the dual-channel photometer (Fig. 4). All hardware was in place and functioning but it should be understood that some of the items from Princeton had been supplanted by later construction and purchases at UP.

Blitzstein continued his collecting of photocells and lab hardware and by then had built an enclosure that he called the banjo from its imaginary resemblance in size and shape to the musical instrument. This testing device contained either of two 20 μCi C^{14} sources behind an enclosing window whose Čerenkov radiation could illuminate the cathode of a photocell under voltage. The anode output was then fed to an electrometer or to a strip chart recorder for archiving.

---

[1] `www.gravic.com/about/RHK-Observational Astronomy-UP/index-html.`

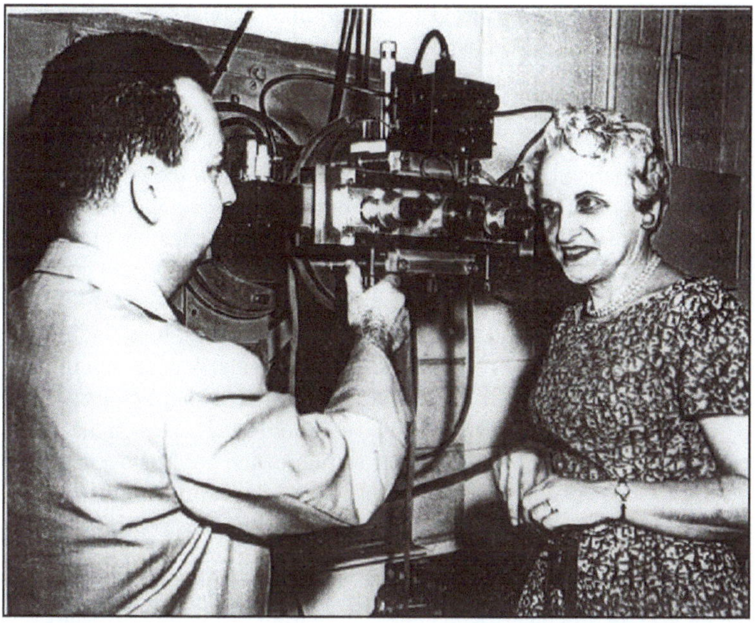

Fig. 4 Blitzstein showing the new photometer to Newton Pierce's widow, Beatrice

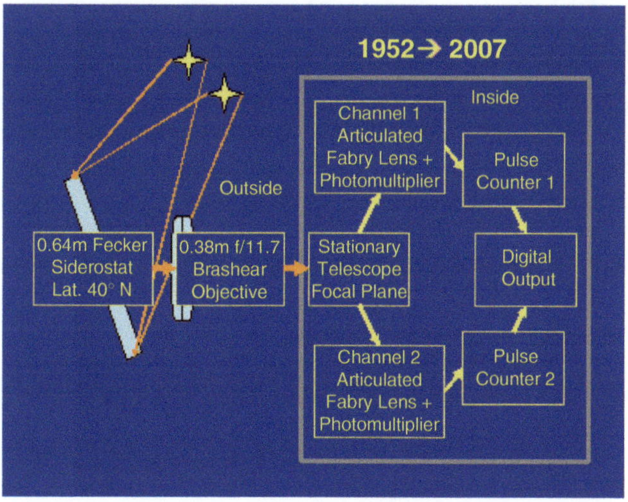

Fig. 5 Schematic light paths from 2 stars simultaneously feeding the photometer

The single-channel description ending the last section may be recollected in order to understand how dual-channel measures were made. Not only did essentially every task already described have to be doubled (Fig. 5), but there were other matters to deal with.

In the first place, both stars had to be acquired optically by the independent setting eyepieces for the two channels. This meant conforming to a decided observing protocol that was never questioned although other choices could have been made. The telescope design caused the focal plane image to rotate with time about the optical axis at a rate that depended on declination and hour angle. Either one could set the comparison star on the optical axis and follow the rotating position of the program star, or one could do the inverse option, or one could set the optical axis on the imaginary bisector of the imaginary small circle connecting the two stars on the sky. Wood and Blitzstein decided that the third possibility was to be used so the two stars had to be displaced equally from the optical axis because it kept physical exertion to a minimum through a night. When the axial wide-field finding eyepiece showed both stars to the observer, this was not particularly challenging, but many programs required a comparison star that was not visible in that eyepiece at the same time as the program one. Blitzstein mounted two 6-in. steel scales on the photometer frame so the observer could refine his/her setting of the guiding eyepiece for each channel. This successive approximation task could be vexing so the scale settings were recorded in order not to have to repeat the entire exercise each time that a particular program star was observed. The precision with which the setting could be done was about $\pm 2''$ – comparable to the size of the seeing disk. The PBPHOT was mounted on a 11-in. diameter sealed circular bearing so that the two stars could be acquired no matter what their polar angle orientation and the entire assembly could be rotated on this bearing to compensate for the field rotation.

By means of the guiding eyepiece the observer viewed each star through its focal-plane diaphragm. These had been machined to be of identical diameters for the two channels and were 4 in number with angular diameters of $69'', 98'', 147''$ and $245''$. These extraordinarily large diameters were enforced by Blitzstein because he didn't trust the drive and tracking capabilities of the telescope and because the off-axis secondary spectrum of the objective was not negligible. Of course, the penalty was increased foreground sky brightness for all measures.

Color filters for the two channels were always cut from the same stock plates and their transmission curves archived in the observatory files. Typically, 4 or 5 different filters were mounted in the PBPHOT at a given time and the observer changed from one filter to another manually in a fraction of a second. There was also a large stock of neutral density filters available to be chosen so that only time-resolved pulses would be counted for each channel and their transmission curves were on file as well. There was not even a necessity that both stars be observed through the same color filter at one time although the record shows no evidence that this was done. Blitzstein bought dozens of multiplier photocells over the years, kept them under voltage in the dark and recorded the spectral responses of all of them. At any time, those that were installed in the PBPHOT were matched in spectral response as closely as possible from the inventory routine support work for the observational program. Graduate students were employed as Research Assistants from departmental or research grant funds to do such tasks as determining spectral responses, writing code, determining the rate of rotation of the field of view and observing. The more exacting work, such as frequent calibration of all the sub-PBPHOT systems and computer subroutines or modifying hardware, fell to the electronics technicians and machinists.

There was also an additional and important concern. In actuality, the observer did not know the absolute or relative responses of the two channels and he did not know whether or not they were constant with time. This detail was critical to obtaining credible light curves. Until 1966 Blitzstein met this problem by installing a ($Sr^{90}+Y^{90}$) source behind a fused quartz window and a blue filter so that the Čerenkov emission from the window was fed to each photocathode in turn without passing through the usual color or neutral filters. The technique could lead to only a single ratio of the channel sensitivities and had to be supplemented by additional color-dependent measures. The radioactive source was initially rated at 100 μCi and the ratio of the Čerenkov counts was defined to be the ratio of the sensitivities of the two channels of the PBPHOT. After about 10 years, radiation damage to the quartz window had become noticeable and Blitzstein designed and built what came to be called the IL, the Incandescent Light Source. It was not unlike an integrating sphere in that a Mazda #51 flashlight bulb was mounted at the end of a truncated pyramidal box painted white on the inside. The lamp illuminated a flashed opal glass window positioned by hand just to the sky side of the telescope objective and filled the objective completely. Before observing could begin, PBPHOT measures of the IL were taken repeatedly and were followed by repeated measures of the same source with no power applied to the lamp. The observer made similar measures to end his observing run on a star and, if he followed Blitzstein's admonitions, he also made similar calibrations during the night too. Because the system sensitivity response depended on the cross-axial positions of the two channels, calibrations always had to be made when changing from one star pair to another.

Lastly, the observer had to be aware of what the telescope was seeing. If it drove too far east or west, it could begin looking at the walls or the south roof and the objective not be illuminated entirely by the stars or sky. In the 1960s Blitzstein set George W. Wolf, then a grad student, to the task of mapping complete sky visibility as a function of hour angle and declination. There was also another option to be weighed – the observer could use any of 11-in., 10-in. or 7-in. objective diaphragms in order to eke out a little more unvignetted sky coverage. This meant, of course, that transmitted stellar flux was diminished, the cathode was illuminated by a smaller light spot and counts became fewer, but this might not be a problem for very bright stars. In fact, the possibility led to one of the very few collaborative efforts between the siderostat and the reflector at the other end of the building. In 1959, there was an international collaborative observing program dedicated to β Lyr. Leendert Binnendijki (1913–1984) observed the binary with the reflector while L. W. Fredrick monitored it with the diaphragmed siderostat and PBPHOT. The former (Binnendijk (1960)), but not the latter, observations were published. Not everyone accepted the arguments of Blitzstein (1953). The most forceful rejoinder to that presentation came from Johnson (1962) who considered the 6 specific advantages that Blitzstein had enumerated for pulse counting procedures. Johnson's main points about discriminating against primary photoelectrons and non-linearity of the output signal for bright stars have to be considered even yet. Although it is true that discriminator levels can indeed throw away low-pulse-height photoelectrons, that concern has little or no effect for the PBPHOT which was designed for the mission of observing variable and comparison stars of comparable brightnesses. The loss of comparable numbers of such photoelectrons from both stars hardly compromises the magnitude difference between them. Even if a suitably bright comparison star was not to be found conveniently, use of neutral density filters on the photon stream for the variable largely obviated the concern. The same practice kept coincidence corrections to the observed pulse train to a low value and the difference between the coincidence corrections to a still smaller value. Johnson also remarked on the inevitable statistics-only basis for the pulse coincidence corrections to the counts. Over the years, Blitzstein and RJM improved the circuitry repeatedly in order to speed up counting capability and routinely measured the system's resolving time so as to know its limitations. Their practices diminished, but did not remove entirely, the second of Johnson's criticisms.

The remaining four advantages claimed by Blitzstein remain true for the time when he wrote them but it is true that charge integration was about to be developed into a confident observing methodology and his claims would be anachronistic in just a few years just as Johnson noted.

7 Maturing the PBPHOT System

A major fraction of Pierce's and Blitzstein's visions had been realized by 1955: simultaneous observing of program and reference stars, matched and calibrated responses for the two optical channels, pulse-counting and amplification as the electronic system, modern multiplier photocells as detectors, and continuing testing and improvement of the system. How well had they done to equalize the airpaths of the two stars?

This question may be examined by two sets of calculations of first-order atmospheric extinction, defined as $\sec z$, the secant of the unrefracted zenith distance angle. We imagine that the two stars are separated by the minimum and maximum angular extents permitted by the design, $0\overset{\circ}{.}1$ and $1\overset{\circ}{.}6$, respectively. The first set of calculations considers the two stars to stand on a unique hour circle at meridian passage and the second set of calculations supposes that they reside on a parallel of declination at the limit of unvignetted visibility just above the east wall of the telescope room. In the senses of northern star *minus* southern one and eastern star *minus* western star, these differences are shown in Fig. 6.

The non-monotonic runs of these functions are due to the ways in which the walls and roof controlled limiting hour angles of visibility. With local extinction coefficients of the order of 0.25 and 0.4 for red and yellow bandpasses, respectively, it can be seen that differences of would be very small over the entire visible sky if the two stars are situated at the smallest angular separation. If, on the other hand, the separation between the stars were as large as the PBPHOT permitted, extinction differences in the second decimal place of a magnitude difference would be common

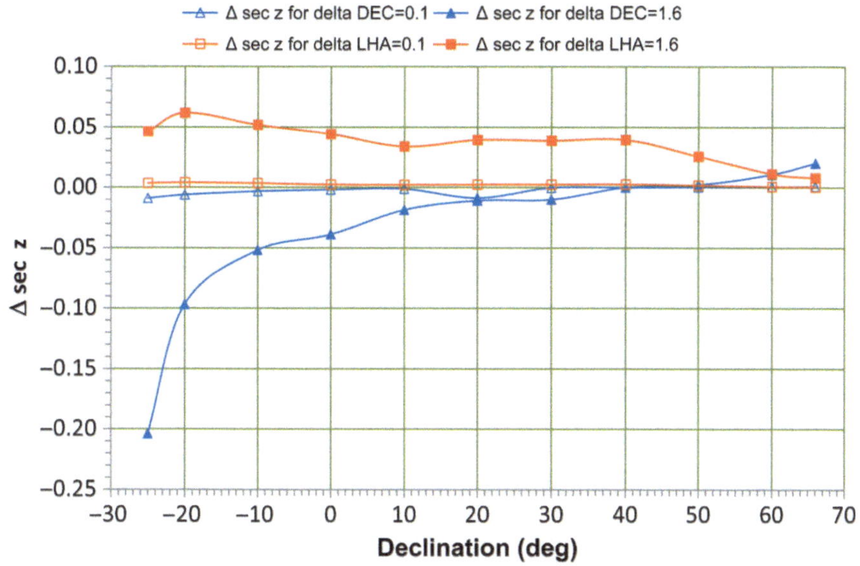

Fig. 6 The minimum and maximum airmass differences for stars in the two PBPHOT channels. The structure of the curves is largely determined by the configuration of the observing room walls

at least at southern declinations. Most star pairs will, of course, have been observed at intermediate separations and orientations of the PBPHOT between the limiting aspects that are the basis of the calculations leading to the display in the figure.

Extinction coefficients need not have been the same for the two stars but could be suited to the filters and color temperatures of the two objects. As a result of this flexibility, Blitzstein sustained an exhaustive study of terrestrial atmospheric models beginning with Blitzstein, Fliegel and Kondo (1970). Largely, because he was convinced that sky brightness levels and their fluctuations would always be the limiting factor in observational precision and also because he had confidence in the discriminator circuit design, Blitzstein never refrigerated the photocells. Repeated requests to do this were turned aside.

It was to be expected that hardware components of the PBPHOT would be continuously replaced and updated. Funds for these changes came from NSF grants to a few people or from observatory budget lines. It is difficult to indicate all small-component changes such as filter additions and replacements and all the interface boards designed, built and installed, but the following is an attempt to note the chronology of on- and off-line sub-system upgradings.

- *Field Rotation Control*: manual control (1948); CY512 stepper motor control and driver (1999).
- *Time Base*: General Radio Vacuum Tube Tuning Fork 816 (1953).
- *Power Supplies*: pack of 20 waxed Burgess XX30P batteries (1948); Fluke HV supply (1965); Power Designs, Inc. 2K10 HV supply (1972); Power Designs, Inc. 2M20 HV supply (1990).

- *Detectors*: RCA 1P21s and 1P28s (1948); matched RCA 1P21s (1955); matched RCA 4509s (1983); matched Hamamatsu R1548s (1996).
- *Pulse Amplifier/Discriminators & Counters*: homemade devices (1948); a second generation of homemade devices (1955); paired Cosmic Radiation Labs 101-A A/D circuits (1961); AMPTEK A101 ICs (1977); Stanford Research Systems SR400 dual photon counter (1988); OPA621 op-amps (1996).
- *Support Computing*: UNIVAC SS-80 using FORTRAN(1962); IBM 1620 using FORTRAN I (1963); IBM 7040 using FORTRAN IV (1964); Ohio Scientific C8PDF mini-computer w/ real-time video display of counts (1980); IBM desktop computer w/ real-time video display of counts (1989).
- *Hard-Copy Recording*: Monroe Listing Machine (1948); Streeter-Amet Printing Recorder (1952); IBM 024 card punch (1966); OSI mini-computer (1986); Epson MX80 (1987).

With the installation of the Stanford SR400 the conceptualization of the system passed from hardware to software and what had been about 11 ft^3 of components dropped to about 2 ft^3. More and more intelligent computer prompts to the observer helped avoid mistakes at the telescope and more operations passed under computer control. When a program was finished for a night, a first reduction was done with the on-line computer in several seconds and paper output – such as the entire chronology of activity, or the reduced data, or both – could be had if it was useful.

Consider the blue measures on the night of December 09–10, 1988. The beginning and ending Channel Ratios indicate a 0.05% decrease of the response of Channel 1 to that of Channel 2, a common value. Typically, there was no way to understand this effect but in the reduction process the CR was linearly interpolated to the time of each dual measure. A quantity SIG represented 1 standard deviation for each of the number of counts, CR and magnitude difference. Both resolution corrections and SIG values were based on theory originally derived for nuclear experiments and are cited in the Blitzstein (1988) paper.

The dataset of the same night permits examining the effect of coincidence corrections on the measures. The first-order approximating equation of Blitzstein's summary was used with the measured resolving time of 1 μs. Because the pulse counts for each channel are very similar to each other and because the correction for pulse coincidences in each channel is of the order of 1%, the systematic effect on a single magnitude difference must be significantly smaller than 1%.

Whether or not the SIG values are errors that realistically represent the astronomical and atmospheric noises of the observations may also be judged by the mean of the 27 dual measures in the same dataset. That mean magnitude difference is -2.2077 ± 0.0016. Were the individual theoretical values of SIG realistic, the error of the mean should be of the order of ± 0.0006. The explanation for the discrepancy between the mean theoretical and observed noise values is to be found in a few causes. One of these is seated in the vagaries of the telescope tracking which always caused the spots of light illuminating the cathodes to wander slightly despite the Fabry lenses feeding those cathodes properly. A second cause is due to scintillation noises uncorrelated between the two channels. With the neutral density filters used on this night, the single count levels fall in the domain where scintillation and shot noises are comparable to each other, each of the order of ± 0.0010. Their sum in quadrature accounts for almost all of the observed noise. We examine the residue of noise in terms of general atmospheric effects, considering the possibility of different numbers of scatterers and discrete atmospheric density packets for the two light paths even though their angular separation was only $0.^{\circ}27$ on this sample night. If, as the observer's remark indicates to be possible, there were fog at about an altitude of 50 feet above the station, it could be very slightly non-uniform over a lateral scale of only 3 in. and cause, or at least contribute to, the observed noise in the nightly mean magnitude difference and its error. (The telescope log actually fails to confirm the presence of fog on this night and other evidence indicates it to be absent also.) Should there have been unnoticed thin cirrus at an altitude of 20,000 feet, water droplet or aerosol inhomogeneities over a lateral distance of only 100 feet could lead to the same excess noise level. From the SIG values that have been quoted above, it can be seen that extinction coefficient variability in the third decimal place would suffice to explain the data noise. Typically, therefore, atmospheric inhomogeneities over the indicated lateral scales could be at the level of a fraction of 1% and all observed noise would be accounted for. Johnson's second critique concerning the use of only statistical criteria for evaluating photometric results is seen to be valid at least as it applies to calculated theoretical errors.

A listing such as the one discussed above (which pertains only to the blue filter data and not those of the other two filters of the night as well) was customarily followed by 4 graphics: (1 and 2) plots of the counts for each star (turned into magnitude scale and named TMAG) as a function of Julian Day Number; (3) a plot of the TMAG values of the comparison star as a function of its airmass in order to check the assigned extinction values; and (4) a plot of DELMAG against the Julian Day Number. For the case of the chosen illustration the $\sec z$ leverage is only about 0.21 and the comparison star counts leads to an extinction coefficient of 0.36 ± 0.02 for the blue filter, an impossibly low value for the elevation of the station above sea level. The same interpretation also applies to the yellow and red observations of the night. Most likely, the extinction was time-variable but the design and operation of the PBPHOT still led to credible magnitude differences for the night – exactly what it was designed to do.

8 Contributions to Astronomy

The PBPHOT system had a functioning lifetime of about 55 years so it must have been among the longest-serving devices in the history of astronomical photometry. For comparison, the polarizing photometer at the PUO was in service 32 years. Some telescope time was used for training which could effectively be done only at night. A number of observers found mastering the system to be laborious compared to the more conventional charge-integration system mounted on the reflector and they were known to express their opinions with some emotion. Other more technically-capable people liked the system. There were three causes for this frustration. The first two were the unconventional telescope design with the siderostat and its enclosing room invisible to the observer and the mounting of the dual-channel photometer itself. This unhandy coupling meant that the observer had to keep 2 polar coordinate and 1 Cartesian coordinate systems in his mind as he acquired the star field and tracked it across the sky. The third cause of annoyance was due to the circumstance that one was not just working with integrated electrical charge or with current but with pulses scaled in volts. This was unfamiliar to observers of only limited E and M lab experience. More than a few potential observers were defeated by the system. When the final instrumental packaging became very compact in its most modernized form, it was essentially a black box and observers need concern themselves with only the few controls available to them.

It was also true that the system could not be efficiently standardized in the photometric sense. If one wanted to do that so as to refer measures to, say, the *UBV* system, one had to acquire standard stars successively in each channel which conceptually defeated the idea of simultaneity. Over a considerable part of the sky there was also not an available range in $\sec z$ large enough to remove extinction effects from such a single-channel effort. Consequently, most measures were left on the natural photometric system.

Published papers, Observatory Reports and the telescope logs testify to an ample diversity of types of observing targets: 7 yellow/red irregular light variables; 4 δ Sct-type, 3 β Cep-type and 2 δ Cep-type pulsators; 12 α^2 CVn-type variables; 17 potential variables in NGC 2264; 96 eclipsing or ellipsoidal binaries including a few new novae; 5 spectroscopic binaries which turned out not to be light variables; 10 apparently single stars that also did not vary; and 2 quasars. Because the PBPHOT had its inception in the PUO polarizing photometer's program dedicated to light curves of variables, it is no surprise that these dominate its history. Because also the staff included Wood, Blitzstein, Merrill and Binnendijk, there is nothing remarkable about the predominance of eclipsing light curves with their potential for basic stellar information. A specific example of this precept may be noted. Whereas most of the binary stars were observed for their complete light curves, a certain number were monitored only for timings of minimum light and a fraction of this number were on the observing program because a general relativistic periastron advance supplements the Newtonian advance.

The total number of dual-channel measures (excluding calibration measures) is estimated to be greater than 250,000 and the logs show that they were accumulated over a very uneven number of nights per year through the lifetime of the instrument. Peak usage occurred in 1978 and 1979 with the system functioning on 118 nights and 110 nights, respectively. These numbers represent nearly 100% usage of suitably clear nights. Thereafter, usage declined raggedly with subsequent peaks of 60 and 39 nights in 1986 and 1999, respectively. Part of this decline was due to the diminution of graduate student interest in light curves generally, part was due to increased usage of KPNO facilities and part was due to one of the busiest observers changing his commitment to the polarization program functioning on the FCO reflector.

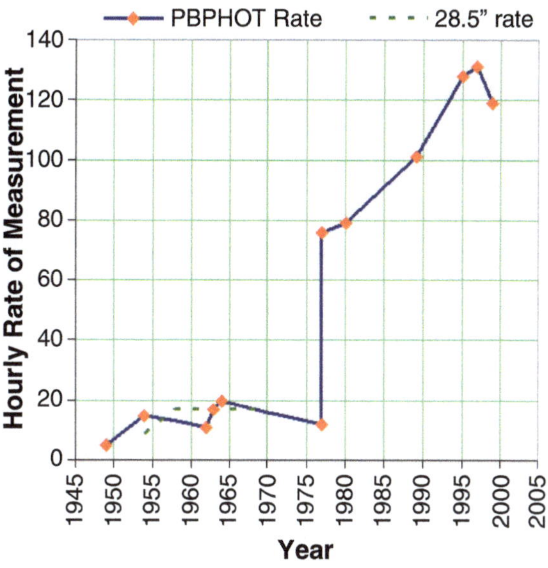

Fig. 7 The chronology of the improvement of the hourly rate of collecting magnitude differences normalized to 20-s counting intervals. The early performance of the single-channel photometer on the FCO reflector is shown toward the bottom left; it did not improve much over its years of service

Until about 1995, there was a relentless urge to improve the functioning of the PBPHOT system by Blitzstein, his technician and some staff and students. This is expressed in the improving efficiency of the sub-systems with IC technology, better photocells, and computer prompting and decision-making to relieve pressure on the observer. For example, it eventually became possible for the computer to drive the entire photometer on its bearing so as to keep up with the field-of-view rotation and thus to free the observer from that chore. A good expression of the increasing efficiency is indicated by the chronology of the normalized hourly rate of measurement and its comparison to the 1950–1960 productivity of the single-channel photometer on the reflector (Fig. 7).

A second kind of program was begun on April 4, 1976 when Mars occulted ε Gem. This event was watched and recorded with the PBPHOT and showed the same range of phenomena as was observed with the Kuiper Airborne Observatory and other North American stations. The FCO must have been very nearly on the central track of the occultation path.

Thereafter, Blitzstein pressed observers to watch for favorable lunar occultations every month and observed some of them himself. The instrument was ideal for this type of phenomenon if a dark-limb immersion was the cause of the occultation. In principle, a reference star could be tracked in channel 2 (the eastern one) while the star to be occulted was isolated in the diaphragm of channel 1 (the western one). With the SR 400, counting times could be as brief as 5 ms and 2,000 counts could be accumulated in the instrument's buffer before they had to be downloaded. When the program terminated, 35 very high-quality occultations had been observed with good immersion timings and some detections of fainter companions. Bright-limb immersions were noisier and emersions were essentially impossible to detect because the telescope did not track well enough to pick up the star after it had become invisible behind the lunar disk.

9 Summary

The argument has been made here that the PBPHOT system and its developmental precursors were the results of several innovative concepts by Newton Pierce and William Blitzstein. As it passed through its successive incarnations, the system became ever better adapted to the eastern U.S. night environment. Useful modernizations marked its development over a half century of productive use. Had the system not been invented, developed and used busily, a significant amount of stellar science would be lacking now and a certain number of productive astronomers might not have been trained as well as they are. The last entry in the log is for June 24/25, 2004.

When the station property was sold to a private individual, there was no opportunity for the staff to remove the PBPHOT system (as well as most of the rest of the installed and stored hardware and papers). The FCO sat idle with no oversight or care by the owner until it was discovered only accidentally that the structure and its contents had been vandalized on more than one occasion. No one was ever charged with the intrusions and damage. At some time, the photometer had been treated with a tool like a baseball bat and there was extensive additional damage to the telescope peripherals. The owner then decided to disencumber himself of the installed equipment and raze the buildings. A deal with a few individuals from The Antique Telescope Society brought all the remaining hardware into their hands, and the siderostat and PBPHOT system reside now at The Florida Community College of Gainesville where there appears to be some prospect of at least the telescope being put into eventual service.

In a way, the eventual fate for the PBPHOT was even timely for CCD cameras would have supplanted it anyway. In the early 1990s two of the authors had imagined that they might replace both photocells with two such chips but always found time to keep busy in other ways. Had they achieved this further advance, there would be even greater cause to regret the loss of this imaginative system.

Acknowledgments We are happy to acknowledge the assistance of Sybil Csigi, Greg Shelton, Librarian of the USNO, and Bruce Holenstein of Gravic, Inc.

References

Binnendijk, L. (1960). *Astronomical Journal*, **65**, 84

Blitzstein, W. (1950). *Astronomical Journal*, **55**, 165

Blitzstein, W. (1954). *Astronomical Journal*, **59**, 251

Blitzstein, W. (1953). In F. B. Wood (ed.), *Astronomical Photoelectric Photometry*, (Washington: American Association for the Advancement of Science), p. 64

Blitzstein, W. (1958). In F. B. Wood (ed.), *The Present and Future of the Telescope of Moderate Size*, (Philadelphia: U. Pennsylvania Press)

Blitzstein, W. (1988). *Vistas in Astronomy*, **32**, 181

Blitzstein, W., Fliegel, H. F., & Kondo, Y. (1970). *Applied Optics*, **9**, 2539

Bouguer, M. P. (1729). *Traité d'optique sur la gradation de la luminière*, (Paris: C. Jombert)

Hearnshaw, J. B. (1996). *The Measurement of Starlight – Two Centuries of Astronomical Photometry*. (Cambridge: The University Press)
Herschel, J. F. W. (1847). *Results of Astronomical Observations Made During the Years 1834, 5, 6, 7, 8 at the Cape of Good Hope*, (London: Smith, Elder & Co.)
Johnson, H. L. (1962). In W. A. Hiltner (ed.), *Astronomical Techniques*, (Chicago: U. Chicago Press), p. 174
Kron, G. E. (1946). *Astrophysical Journal*, **103**, 326
Levitt, I. M., & Blitzstein, W. (1947). *The Pennsylvania Gazette*, **46**(2), 12
Pickering, E. C. (1912). "Observations with the Meridian Photometer During the Years 1902 to 1906," *Annals of the Astronomical Observatory of Harvard College*, **64**, 1
Pierce, N. L. (1947). *Copris Pristinus*, 22
Wood, F. B. (1946). *Copris Pristinus*, **21**, 1
Wood, F. B. (1951). *Copris Pristinus*, 25
Wood, F. B. (1952). *Pennsylvania Gazette*, **50**(9), 14
Wood, E. B., & Blitzstein, W. (1957). *Astronomical Journal*, **62**, 165
Yates, G. G. (1948). *Monthly Notices of the Royal Astronomical Society,* **108**, 476

Johnson Photometry and Its Descendants

Arlo U. Landolt

1 Histories of Photometry

Two sources which discuss the history of astronomical photometry are recommended to the reader: (1) a series of papers in the defunct serial Popular Astronomy [QB1.P8] by Weaver (1946a–1946f), and (2) the eminently readable book The Measurement of Starlight: Two Centuries of Astronomical Photometry by Hearnshaw (1996).

Weaver divided photometric history into four time periods. The initial period involved the human eye. This period was the time in history when observers divided the naked eye stars into six brightness intervals, labeled magnitudes. Accuracies of eye-estimates probably were no better than 0.25 magnitude, and some sources suggest 0.4 magnitudes as more appropriate.

Weaver's second period saw the use of mechanical instruments, such as polarizing photometers and meridian photometers. The biggest advance, though, was the introduction of a standard brightness scale, the Pogson scale, set by the statement that a light flux ratio of 100 is equivalent to a difference of exactly five magnitudes in brightness.

Weaver's third period began with the introduction of photography, and its application to astronomical photometric problems. The Moon was recorded in 1839, and by 1850, Bond (1850) at Harvard was able to image stars via the photographic process. One of the first star catalogs based on photographic magnitudes was published by Pickering (1890).

Weaver's fourth period (Weaver 1946e), continuing to modern times, saw the use of physical instruments. Insofar as we are concerned, this period of astronomical photometry began with Stebbins' observations with a selenium cell in 1907 (Stebbins and Brown 1907). Photoelectric photometry gained additional attention later when Bond (1946) showed that the characteristics of the RCA 1P21

A.U. Landolt (✉)
Department of Physics and Astronomy, Louisiana State University, Baton Rouge, LA 70803, USA
e-mail: landolt@phys.lsu.edu

E.F. Milone and C. Sterken (eds.), *Astronomical Photometry: Past, Present, and Future*, Astrophysics and Space Science Library 373, DOI 10.1007/978-1-4419-8050-2_5,

photomultiplier was 10–15 times more sensitive than any other device previously used for astronomical photometry.

A short summary of the preceding paragraphs, with the addition of more recent astronomical photometric efforts, may be found in Landolt (2007).

2 Definitions

The astronomical literature utilizes a number of terms which describe the kind of photometry undertaken for a given project. A description of the terminology is in order.

Relative photometry is the kind of photometry which most observers do in their studies. It is photometry tied into sets of standard stars established around the sky, with zero points which can be traced back through photometric history. Such measurements are not tied into any laboratory system, but are related to nearby standard stars, in a variety of photometric standard systems.

Absolute photometry is based on spectrophotometry, or photometry tied into a laboratory source, into a black body cavity, or something similar, all as an integral part of the data acquisition process. Absolute photometry is based on physical units. In spite of the terminology used in much of the recent literature, only a small number of astronomers (e.g., Art Code, James Gunn, Bev Oke) ever have done absolute photometry.

Differential photometry is the direct comparison of two or more stellar images, historically best done via a photographic plate, but now via CCD imaging. Many stellar images are obtained on the same photograph, or CCD image, and hence can be measured, inter-compared, with high precision because the air masses essentially are identical. One directly compares the intensities of two nearby images, determining the difference in magnitudes, and perhaps then plotting the result versus time to search for a light variation of the object under study. Examples chosen at random include a long series of eclipsing binary system light curves by Johannes Andersen and colleagues (e.g., Andersen et al. 1984), a similar series of light curves on intrinsic variable stars by D. H. McNamara and colleagues (e.g., Alexander et al. 1987), and the realization that one of his potential standard stars was variable by Landolt (1990).

An act of measurement provides a datum which has associated with it a number indicating the error involved in making that measurement. One can define two related errors when doing photometry, one describing the error in making the actual measurement with the instrumental set-up at hand, and the second the error associated with the measurement after that measurement has been transformed to a standard photometric system. The first error is called the internal precision, and indicates the error obtainable after repeating the measurement again and again; internal precision describes the repeatability of the measures. The second error is called the external accuracy, and results from comparing measures of the same object again and again once those instrumental measures have been transformed to a standard photometric system. Depending on the manner in which the transformed measures are analyzed, one can calculate a standard error of a single observation, a mean error of the mean observation, and so on. The essential difference is that precision refers to a discussion of the measurements as made with a particular instrumental set-up or arrangement, and accuracy refers to a discussion of data after they have been transformed from a particular instrumental system to a standard photometric system, thereby allowing the intercomparison of different observers' photometric results.

3 The Lineage of the *V* Magnitude

There always has been a common thread through observational astronomical history, tying together the different photometric systems (see Landolt 2007, Sect. 2). The chronological thread tying together historical photometries up to the late twentieth century, was the human eye (effective wavelength about 5,500Å = 550 nm), through the late nineteenth century, the photovisual magnitude via photography from the 1890s to the 1970s, the *V* magnitude (effective wavelength 5,550Å = 555 nm) in the *UBV* photometric system from the early 1950s to the present, the *y* magnitude in the Strömgren *uvby* four-color system from the 1960s to the present, the *V* magnitude in the Geneva system from the early 1970s to the present, the *V* magnitude in the Walraven system from the mid-1970s, and the *V* magnitude in the Vilnius system from the early 1970s (see Drilling and Landolt 1999, Table 15.5).

One could carefully track through Weaver's papers (Weaver 1946a–f) and Hearnshaw's book (Hearnshaw (1996)) a complete history of the visual magnitude, *V*. Suffice it to say here that the visual magnitude, *V*, in use today, has a zero point adjusted to agree with the magnitudes of the North Polar Sequence stars given by Stebbins et al. (1950).

4 Photographic Photometry

A single great advantage of photographic plates is the ability to record data for many stars during one exposure, but with the cost of lower accuracy. On the other hand, a photoelectric photometer can observe only one star at a time, but with an increase of four to five times in accuracy. In some ways, the photographic plate remains the best medium for long-term data storage. It has been demonstrated that photographic plates can be stored for more than a century, with their information content still adequately retrievable. A down side is the effort needed to retrieve information, to measure the images, and then turn this information into magnitudes and color indices.

Errors of photographic plate measures strongly depend upon the quality of the image. For round, sharp, well-exposed images about three magnitudes above the plate limit, the mean error of a photometer setting for Cuffey's iris photometer was about a percent (Cuffey 1956). The plate error, though, which arises from the characteristics of the photographic process, was much larger than the measuring error. One could expect the mean error of a single measurement of a star image to lie in the range of 0.03–0.04 magnitudes.

Weaver (1962) wrote that for a star image well above the photographic limit, a mean error for a single measurement of a star image from good quality, large reflector photographs could be expected to lie in the range of 0.03–0.04 magnitudes.

This author Landolt (1964a,b) did a photoelectrically calibrated photographic study of two open clusters which contained classical Cepheids. The photographic data were *V* and *B* plates taken at the Royal Cape Observatory Victoria refractor. That telescope has rigidly connected 24-in. photographic and 18-in. visual lenses. The errors of a transformed mean photographic magnitude for the linear portion of the iris photometer reduction curve were in the range of 0.04–0.05 magnitudes. Remember that these were refractor data.

The very best photographic photometry, reduced with extreme attention to detail, probably achieved a standard error of 0.02 magnitudes.

5 The Introduction of the 1P21 Photomultiplier

The leaders in the introduction and early use of photoelectric photometry in the modern era, post-World War II, were Gerald E. Kron (1913–), Olin J. Eggen (1919–1998), and Harold L. Johnson (1921–1980).

A perusal of the literature indicates that Kron already had been doing photoelectric photometry since the late 1930s (Kron 1938, 1939). After his return from wartime jobs, he began to use newly available technology, the 1P21 photomultiplier made by the Radio Corporation of America (RCA), and DC amplifiers which he had helped design. Data were recorded with strip chart recorders.

Kron (1946) reported on his initial use of an 1P21, first available in 1943, and discussed its characteristics. The 1P21 featured a blue-sensitive photocathode made of layers of cesium and antimony (CsSb). Whitford (1977) wrote that the 1P21's success was based on three facts: (1) it had a much higher quantum response due to the cesium antimony photocathode, (2) it had a much wider spectral response than had the early photo-cells, and (3) the multiplier process provided a close to noise-free system of amplification.

One of the first users of the 1P21 was Olin Eggen, who worked both at the Washburn and Lick Observatories with equipment designed in great part by Kron. Eggen (1950a,b,c) did early photoelectric studies of the color magnitude diagrams of the Hyades, Pleiades, Coma Berenices, and Ursa Major galactic, or open, star clusters. This work was carried out at refracting telescopes, using blue and yellow filters which were close to the peak sensitivities of the blue-sensitive and yellow-sensitive photographic emulsions. These data were tied into North Polar Sequence stars. Other photometrists used similar equipment arrangements. Kron proposed that the photomultiplier based results be called the *P*, *V* photometric system (Kron and Smith 1951).

Photometrists soon found that the magnitudes and colors of stars depended upon the mix of the North Polar Sequence stars chosen to be used as standards (Stebbins et al. 1950). Hence, improvements in the photometric system were needed.

6 Introduction of the *UBV* Photometric System

The introduction of photoelectric photometry brought with the technique a number of advantages over the century old process of photographic photometry. Photomultipliers featured high dynamic range, a much better quantum efficiency, and a linear response to the incoming radiation. Furthermore, the data reduction process was straightforward. The major disadvantage was that only one star could be measured at a time, whereas the photographic plate had the capability of recording large numbers of celestial images with one exposure. On the other hand, accuracies achieved via photoelectric photometry were five to ten times better than those derived from photographic photometry, mostly dependent upon the care and attention to detail, practiced by the astronomer.

The decision to so define the zero point of the *V* magnitude was taken at a meeting on July 6, 1950 in Pasadena, California. Attendees at the meeting included Bowen, Baade, Baum, Minkowski and Pettit of the Mt. Wilson and Palomar Observatories; Eggen, Kron, and Weaver of Lick Observatory; Whitford of the Washburn Observatory; and Johnson of Yerkes and McDonald Observatories (Johnson and Morgan 1951).

The famous paper of Johnson and Morgan (1953) initially defined, illustrated, and provided stars for the use of the *UBV* photometric system. A RCA 1P21 photomultiplier served as the detector. A year later, the definitive paper appeared, providing a list of 108 standard stars around the sky (Johnson and Harris 1954). Johnson later expanded the discussion (Johnson 1955) and in Johnson (1963). In the latter paper, he discussed photometric systems as defined by photoelectric techniques.

The published standard values for the *UBV* system were exactly the quantities (in terms of magnitudes) that had been observed. That meant that the *UBV* photometric system is the natural system of the original apparatus. The *V* magnitudes and the $B-V$ and $U-B$ color indices are on Johnson's original instrumental system. The zero points of the *UBV* color indices, $B-V$, and $U-B$, were defined to be zero, were set to zero, for A0V stars (Johnson 1963).

Johnson pointed out that reflecting telescopes with aluminized mirrors should be used if at all possible. Use of reflecting telescopes whose mirrors had silver coatings, or use of refractors would introduce non-linear transformations, particularly for the *U* filter.

7 Photographic Photometry with the *UBV* System

The quarter century 1950–1975, roughly, saw astronomical photometry based on the direct photoelectric calibrations of stellar sequences recorded on photographic plates, emulsions. By 1950, when the *UBV* photometric system came along, photographic photometry already had been underway for 50 odd years. Johnson followed up his and Harris' paper which had defined 108 standard stars around the sky with his paper in the Annales d'Astrophysique (Johnson 1955). Johnson described in this paper the inadequacies of the International System of Magnitudes and of the North Polar Sequence when one got to the point of standardizing accurate photometry. He again listed the new *UBV* photometric standard stars. In the Annales d'Astrophysique paper, Johnson also laid out the best filter prescription for those observers wanting to do optimum photographic photometry on the *UBV* photometric system. His filter prescriptions were:

- $U =$ Corning 9863, or 2 mm Schott UG2 + blue-sensitive emulsion
- $B =$ Corning 5030 + 2 mm Schott GG13, or 1 mm Schott BG12 + 2 mm Schott GG13 + blue-sensitive emulsion
- $V =$ 2mm Schott GG11 + yellow-sensitive emulsion

The photographic emulsions used were manufactured by Eastman Kodak. The blue-sensitive emulsions were 103a-O, or IIa-O. The yellow-sensitive emulsions either were 103a-D, or IIa-D. The IIa-O and IIa-D emulsions essentially had the same effective wavelength sensitivity, but were finer grained emulsions, than were the corresponding 103a-O and 103a-D emulsions.

8 Addition of Photometry at *R,I* Wavelengths

Prior to Johnson's extensive efforts at photometry in the long wavelength end of the visual spectrum, Kron had established a photoelectrically defined *R*,*I* photometric system. Kron and Smith (1951) observed 125 stars in ten areas around the sky using a Continental electric type CE25A/B photocell. The effective wavelengths of their *R* and *I* filters were 6,800 and 8,250Å with full width at half maximum of 1,850 and 1,480Å. Kron, White, and Gascoigne (Kron et al. 1953) observed 138 stars for use as standard stars, obtaining their data at Mt. Stromlo in Australia. Their accuracies appeared to be at the two to three percent level.

Johnson (1964) later extended the *UBV* system to longer wavelengths via *R* and *I* filters. In those days it was necessary to use a different photomultiplier for the *R*,*I* filters, than was used to do *UBV* photometry. One should note that the effective wavelengths of the Johnson *R*,*I* filters were 7,000 and 8,800Å, with half widths of 2,200 and 2,400Å, respectively.

Johnson stated that the *R*,*I* photometry published for stars in Johnson (1964) "should be regarded as the standard values for additional photometric work." A recasting of the same material, which appeared in Iriarte (1965), indicated that the data had been obtained with a RCA 7102 photomultiplier tube. This paper provided probable errors of a single observation for the *V* magnitude and the various colors, on the order of, or just under two percent.

Even though standard stars have been made available for Johnson's version of a *UBVRI* photometric system (Moffett and Barnes 1979a,b), the community's use of Johnson *UBVRI* has declined with time. A reason may be that fainter standard stars on the Johnson system were published after Cousins–*RI*-based photometry already was in general use. Another reason may be that because the Johnson *R*,*I* filter effective wavelengths were 600–900Å farther to the red, to longer wavelengths, atmospheric effects caused more problems with the data, hence leading to lower accuracy. Most of the data were obtained with an extended-red S-20 instrumental system.

Cousins (1976) introduced use of the RCA 31034A photomultiplier to the astronomical community via the publication of standard stars to be used with it. That photomultiplier's gallium arsenide photocathode provided still greater sensitivity, and even more important, a greater spectral sensitivity. Now for the first time, astronomers could gather *UBVRI* data using one detector, the RCA 31034A. It should be noted that as Cousins was observing and accumulating data with the RCA 31034A photomultiplier, Weistrop (1975) already had shown that measures made with that brand photomultiplier satisfactorily transferred those data to the Kron *R*,*I* standard stars.

Considerable effort has gone into summarizing the errors identified by Cousins in his many photometry papers and in his published standard star lists. There seems to be no one place where he summarizes his errors. One can only say that his results are exquisite, with standard errors on the order of or less than 0.005 magnitude.

Because Cousins' standard stars tended to be bright, Graham (1982) brought into use at CTIO a RCA 31034A and appropriate filters. He published a set of fainter standard stars around the sky in the Harvard E-regions, centered at -45° declination. All of the author's *UBVRI* standardization work at CTIO has been based on Graham's implementation of the *UBVRI* photometric setup at CTIO, and the author thanks him for his support and counsel over the decades.

Finally, in any discussion of the implementation of the *UBVRI* photometric system, mention must be made of the defining work of Bessell (1976, 1979). He definitively showed the usefulness of the GaAs photocathode, comparing results from it with other photometric systems then in use. Bessell also discussed filter combinations which would be optimum for use with the RCA 31034A photomultiplier.

A summary of the effective wavelengths for different *R*,*I* filters is shown in Table 1. More details may be found in Drilling and Landolt (1999).

Table 1 *R*,*I* filter effective wavelengths λ and halfwidths *W* in Angstroms

Filter	λ_{Kron}	W_{Kron}	$\lambda_{\mathrm{Johnson}}$	W_{Johnson}	$\lambda_{\mathrm{Cousins}}$	W_{Cousins}
R	6800	1850	7000	2200	6400	1750
I	8250	1480	8800	2400	7900	1400

9 Photometry with CCDs

During the past 15–20 years, charge-coupled devices, CCDs, have become the detector of choice for most astronomical imaging programs. These devices can record many stars at a time with exquisite precision, and with accuracies approaching or equivalent to those from photoelectric photometry. As with many, many instances in life, final results from such data depend upon the observer's attention to detail, and the care exercised in the observing process.

CCD detectors have the advantages of high dynamic range, high quantum efficiency, a linear response, and are two dimensional detectors. Another big advantage is the acquisition of star and sky-background measurements at the same time.

CCD detectors have a number of disadvantages: sometimes aggravatingly long readout, or "dead" times; data reduction requires a large effort; shutter-time corrections must be taken into account; cosmic rays must be removed from the image; and, bias, dark and flatfield images must be acquired.

One also should note that most times, CCD data are reduced one filter at a time, in contrast to the usual practice with photoelectric data. With the photoelectric data, data leading to color indices are reduced as a ratio of the counts through two individual filters. Many problems drop out via the ratio approach. The derivation of color indices from two individually determined CCD filter magnitudes leads to larger color index errors, since the error is in quadrature.

There is no doubt but that CCDs will be the detectors of choice of the foreseeable future. Users still await, though, the technology which will assure the longevity of data, as evidenced by well kept photographic emulsions. Because there is no comparison between the final accuracy obtainable with CCD data over photographic data, astronomical users must fervently hope for stable storage possibilities, devices stable for a century, or more.

10 Observing Techniques

A photometric night is one which is clear all, or the majority of, the night. Intermittent atmospheric problems, cirrus for example, a multitude of airplane contrails, or varying haze layers, make a night marginal or unusable. A kind of photometry can be accomplished, but not standardized or standardization photometry.

This chapter is not the place to review a list of positive and negative procedures in doing good photometry, the kind of effort which will lead to the highest accuracy results. Such a discussion may be found in Landolt (2007). Suffice it to say here that there must be sufficient time to obtain measures of both extinction and standard stars every night. Something between one quarter and one-third of a night goes to 'overhead,' that is, to extinction and standard star measurements. Always begin and end

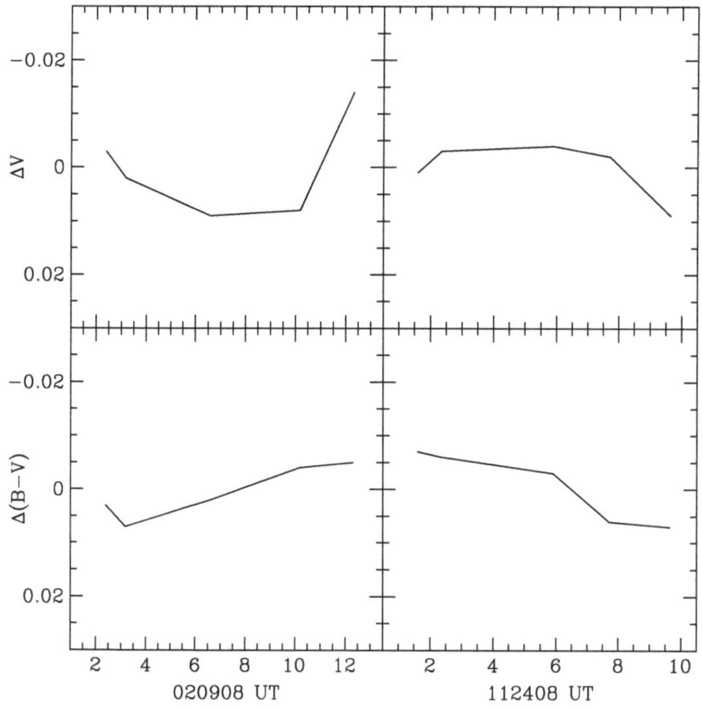

Fig. 1 A plot of the deviations in average V magnitude and $B-V$ color index as a function of Universal Time for the nights indicated. The error of the ordinate typically is of the order of one percent (0.01 mag)

a night's observing with several standard stars possessing as wide a range in color index as possible. Intersperse, every two hours or so, the program star acquisition with a set of standard stars.

The night sky can and does change with time. Most of this variation is due to atmospheric extinction. An illustration is given in Fig. 8 in Landolt (2007). More recent data from two separate nights taken at Lowell Observatory are plotted herein in Fig. 1. In the top part of each night's figure, the ordinate plots ΔV, the average difference between the observed V magnitudes of a group of standard stars and their published values versus Universal Time. Five such sets of stars were observed each night, 9 February 2008 UT and 24 November 2008 UT, over approximately a 10 h interval. Note the systematic changes, changes that must be removed from the program star's magnitudes, if one is to achieve the best accuracy. The bottom section of the plot for each night's data in Fig. 1 provides a similar plot for the $(B-V)$ color index. The reader will note that the variations noted in the figures, both herein and in Landolt (2007) are small, but the point is that they are systematic. Such variations must be taken into account, during data reductions, if one is to achieve the best photometric results.

11 Extinction Behavior Over Time at a Given Location

The behavior of the long term average extinction coefficients over time at a given geographic location is remarkably constant. However, changes from night to night, and within a night, can be dramatic (see Tables 2, 3, and particularly 4 in Landolt (2007). It follows, therefore, that one cannot just blithely use mean extinction coefficients to reduce a night's data, not if one is after photometry of high systematic accuracy.

Long term extinction data for KPNO, CTIO and Lowell Observatory have been tabulated in Table 2. These average long term extinction coefficients are not strictly comparable, since they were taken over different time intervals. However, the mountain sites all have characteristics in common in that they are at similar altitudes, more or less 6,900 feet, and are in semi-arid geographic locations.

General comments involving photometric data reduction may be found in Landolt (2007), with more detailed discussions in books by Henden and Kaitchuk (1982) and Sterken and Manfroid (1992), among others.

12 Photoelectric Photometric Accuracies Over the Years

The entire rationale for the discussion in this chapter to this point is to give the reader an understanding of what is involved in determining quality magnitudes and color indices for celestial objects. The following paragraphs give an indication of the accuracies achieved over the past 40 years or so. Because this chapter is focused on Johnson's *UBV* photometric system, its history, and its descendants, other highly accurate photometric systems are not discussed.

Harold Johnson did not summarize his errors, at least that this author could find. Based on the long term use of Johnson's basic 108 *UBV* standard stars, i.e., based on the recovery of those standard stars' magnitudes and color indices during data

Table 2 Extinction behavior at different sites

	Coefficient	Average Coefficient Values		
	Symbol	KPNO	CTIO	Lowell
V	Q_y	+0.162	+0.152	+0.157
$B-V$	k_1	+0.102	+0.124	+0.135
	k_2	−0.021	−0.023	−0.028
$U-B$	k_3	+0.322	+0.315	+0.330
	k_4	−0.017	−0.022	−0.006
$V-R$	k_5	+0.040	+0.044	+0.049
	k_6	+0.001	+0.007	+0.008
$R-I$	k_7	+0.042	+0.045	+0.035
	k_8	−0.007	−0.006	+0.003
$V-I$	k_9	+0.085	+0.091	+0.082
	k_{10}	−0.006	+0.003	+0.004

Table 3 Graham's E-region standard star accuracies

Range in magnitude/color	Number of stars	Weighted errors
$6.27 < V < 16.7$	100	0.0073
$-0.29 < B-V < +1.91$	102	0.0065
$-1.17 < U-B < +2.15$	100	0.0110
$-0.12 < V-R < +1.12$	102	0.0054
$-0.15 < R-I < +1.32$	99	0.0056

Table 4 Landolt's photoelectric photometric accuracies

	Mean Errors of a Single Obs.				Mean Errors of the Mean			
	1973	1983	1992	2009	1973	1983	1992	2009
V	0.015	0.013	0.016	0.014	0.005	0.003	0.004	0.004
$B-V$	0.016	0.012	0.020	0.019	0.005	0.003	0.005	0.005
$U-B$	0.025	0.023	0.044	0.049	0.008	0.005	0.013	0.014
$V-R$		0.009	0.013	0.012		0.002	0.003	0.003
$R-I$		0.010	0.018	0.017		0.002	0.004	0.004
$V-I$		0.012	0.023	0.021		0.003	0.006	0.005

reduction, one can say that the accuracy of the photometry of those 108 standard stars is between 0.015 and 0.020 magnitudes.

Cousins' photoelectric photometric accuracies are on the order of 0.005 magnitudes, or better. He always quoted errors, but again this author failed to find a location where he summarized them. Suffice it to say that his photometry was exquisite.

The accuracy of Graham's E-region standard stars (Graham 1982) were derived from information in his paper. The derived information is presented in Table 3. The numbers in the last column are a weighted average, based on all stars published in Graham (1982). His stars were observed an average seven times each.

Table 4 lists errors in a series of standard star papers by Landolt (1973, 1983, 1992, 2009). Columns two through five list mean errors for a single observation. The last four columns list the mean errors of the mean magnitude or color index. Recent errors for the $(U-B)$ color index are large since an attempt was made to include stars whose U signal was weak, i.e., the stars were faint in the U band.

13 Comparison of CCD and Photoelectric Photometry

The next paragraphs will illustrate a direct comparison of photoelectric and charge-coupled device (CCD) data for the same sets of stars.

Following the retirement of photoelectric photometers at KPNO, the staff put together an instrumental setup called "CCDPhot." As taken from KPNO's Newsletter No. 43, the CCDPhot system was a CCD-based photometer intended to replace the Mark III "photoelectric photometer that had been available on the KPNO 1.3-m telescope. The CCDPhot setup consisted of a thinned back-side-illuminated Tektronix CCD (designated T5HA) with 512×512 27-μm pixels producing an image scale of $0\overset{''}{.}77$/pixel and a field of view of $6\overset{'}{.}6 \times 6\overset{'}{.}6$ at the $f/7.5$ focus of the KPNO 0.9-m telescope. The CCDPhot software was an IRAF program designed to do multi-aperture, multi-filter stellar photometry in real time. CCDPhot performed all of the functions of a conventional photoelectric photometer, using a CCD in place of a photomultiplier, and software apertures in place of the aperture wheel."

The CCDPhot procedures subtracted the DC offset using the overscan region, subtracted a user-derived bias frame, and divided the resultant image by a user-derived flat-field exposure for the appropriate filter. CCDPhot then provided an instrumental magnitude in each user-defined aperture, subtracted the sky contribution, and wrote the result into a text file. The instrumental magnitudes then could be converted to individual stellar magnitudes and color indices via a normal reduction process.

A set of standard stars was observed with both a RCA 31034A-02 photomultiplier, and with the CCDPhot Tektronix CCD. Similarly, a set of program stars was observed with both instrumental setups. One can see in Figs. 2–4 that all data from the two detectors involving the *BVRI* filters compare to within one percent. Data taken with the *U* filter are considerably worse, perhaps on the order of four percent. Once one learned the system, CCDphot was a really 'nice' observing setup.

The ordinates in the following figures, Figs. 2–4, are differences in the sense CCDPhot results minus photoelectric results. The abscissae all are photoelectric results. The filled circles indicate stars which have been observed five or more times each. The open circles indicate stars observed four or fewer times each. In each slide, standard stars appear in the upper panel, and program stars in the lower panel. The program stars are part of a long term ongoing project. These latter data shown in the lower portions of Figs. 2–4 show considerable scatter since there are many fewer observations per star, and since the program stars are fainter than the standard stars.

A summary is provided in Table 5 of the external accuracies achieved by photometrists over the ages, and more particularly via application of Johnson's *UBV* photometric system. The column headings are rather self evident. The first column identifies a detector type, beginning with the human eye. The last item in the first column is not a detector, but is a technique. Greater accuracies may be achieved by differencing the intensity of one celestial image with that of another (or, even of several) celestial image, a technique called, by some, ensemble photometry. The second column indicates the accuracy achievable via the detector listed in the first column. These accuracy values are meant to be conservative estimates; there always are those few observers who can achieve superb results, results beyond those obtained by the majority of us mere mortals.

The CCD accuracy of 0.005 magnitudes is no smaller than is listed because this is on the order of the accuracy of the final magnitude after transformation. The internal precision of data taken with CCDs certainly can be a factor ten better.

The last column in Table 5 identifies the approximate time frame during which the detectors in column one were in heaviest use. Virtually all the techniques remain in use, although few observers take new data with photographic emulsions. A number of astronomers still do harvest precious photometry from several of the maintained photographic plate archives.

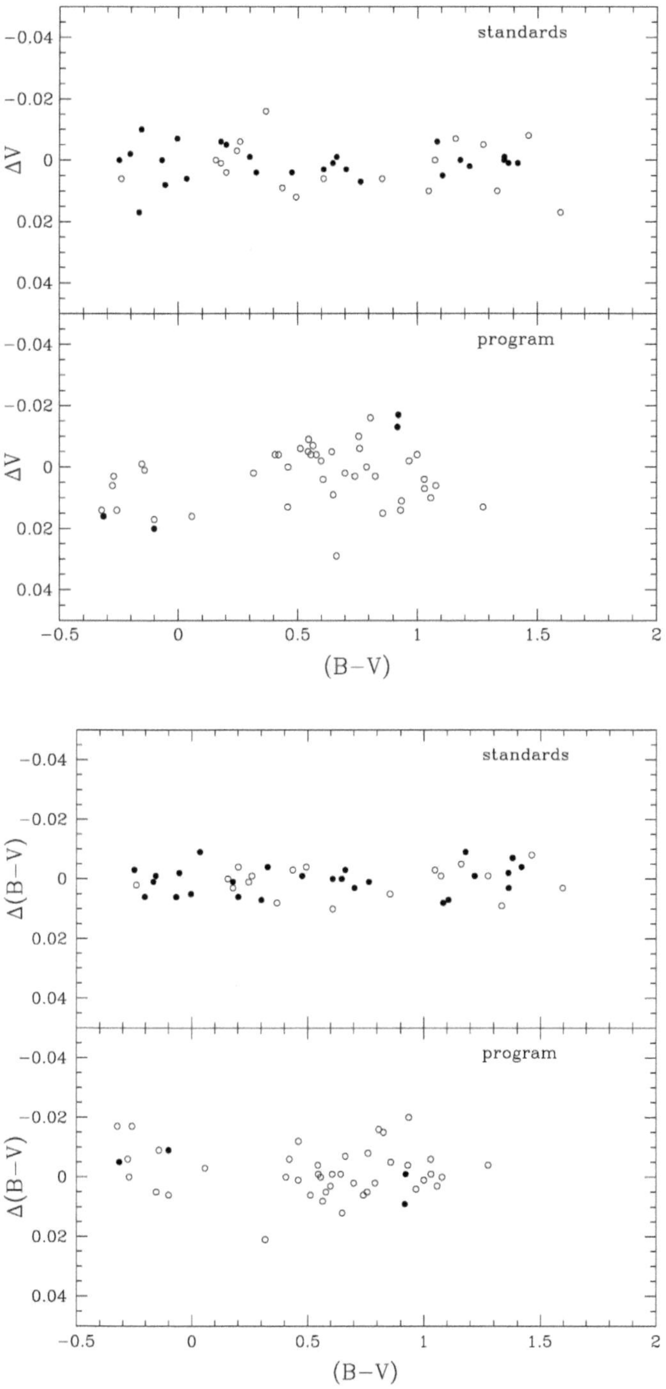

Fig. 2 Comparison of the difference between CCDphot and photoelectric derived V magnitudes (*top*) and $B-V$ color indices (*bottom*) as a function of photoelectric $B-V$ color indices

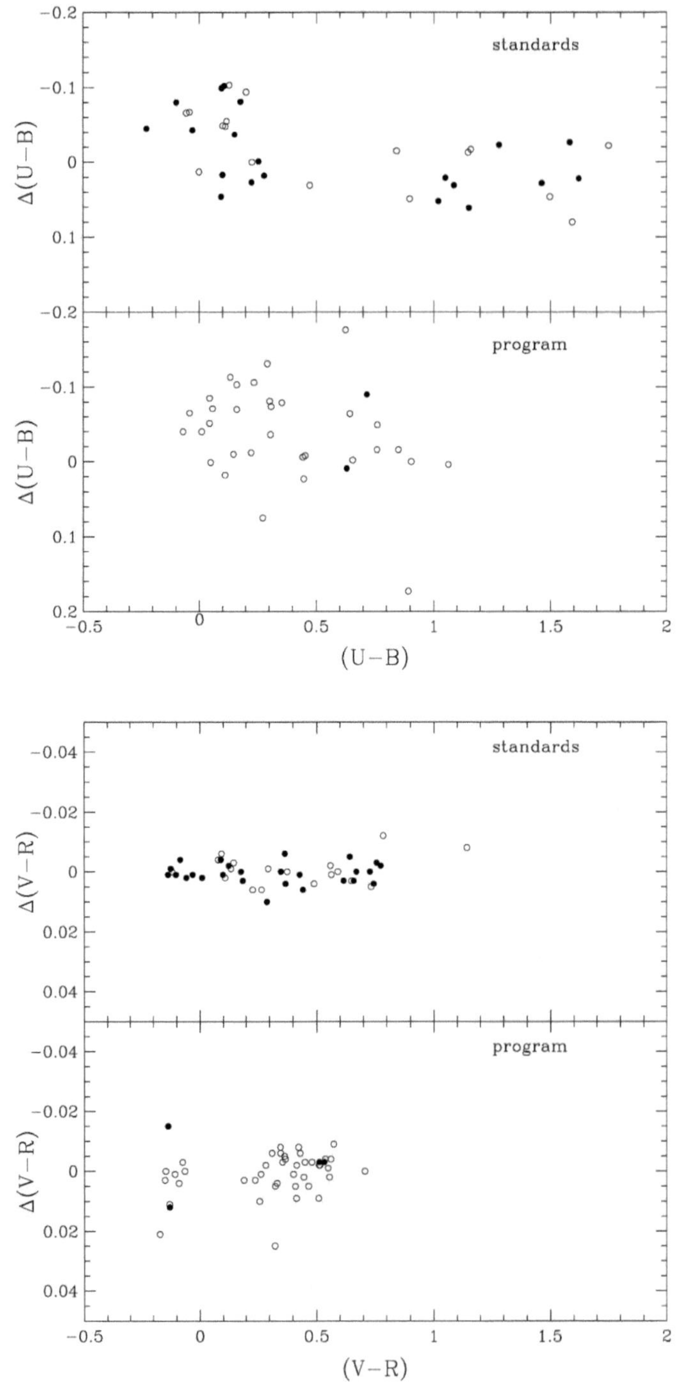

Fig. 3 Comparison of the difference between CCDphot and photoelectric derived color indices: $U-B$ as a function of photoelectric $U-B$ (*top*) and $V-R$ as a function of photoelectric $V-R$ (*bottom*)

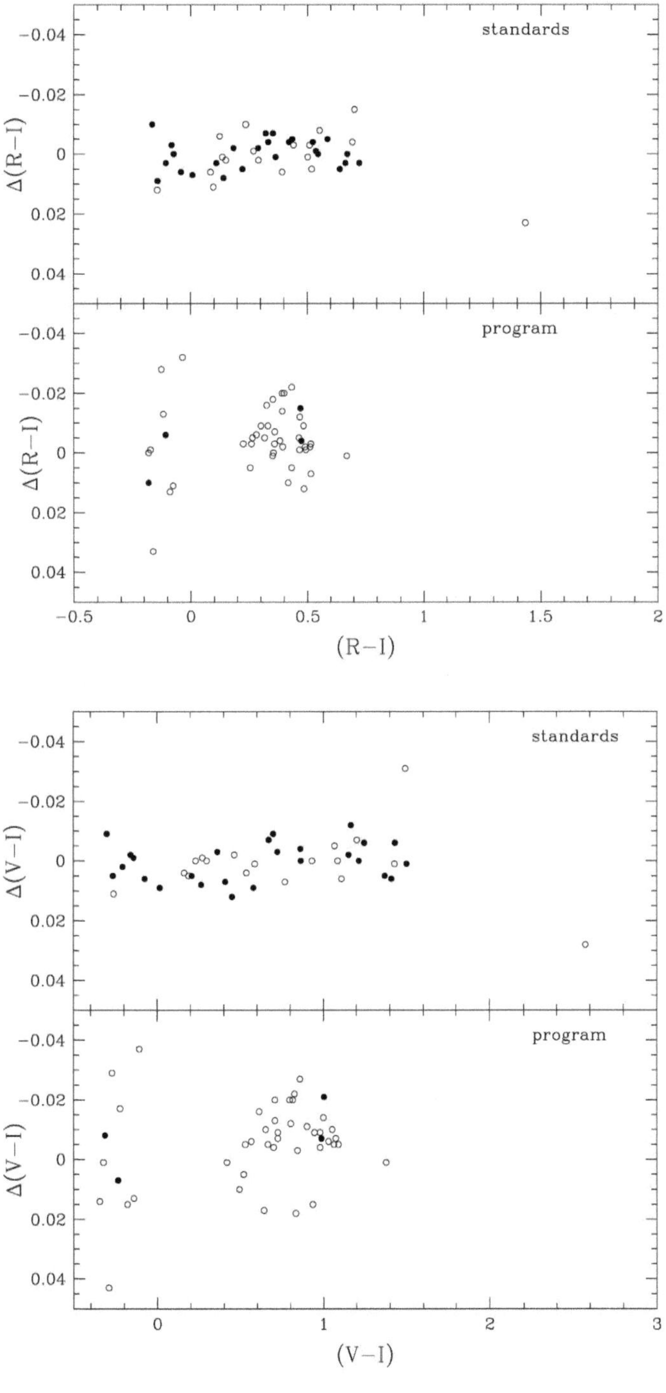

Fig. 4 Comparison of the difference between CCDphot and photoelectric derived color indices: $R-I$ as a function of photoelectric $R-I$ (*top*) and $V-I$ as a function of photoelectric $V-I$ (*bottom*)

Table 5 Summary of external accuracies

Detector	Accuracy	Time Frame
human eye	0.250	~128 B.C. – present
photographic	0.020	~1890 ~ 1970
photoelectric	0.005	~1950 – present
CCD	0.005	~1985 – present
differential	0.0005	

Acknowledgements The author extends thanks, in alphabetical order, to several individuals with whom he has conferred: James L. Clem, David L. Crawford, Arne A. Henden, G. W. Lockwood, Phil Massey, Brad Schaefer, Brian Skiff, and Chris Sterken.

The author's photometric work most recently has been funded by NSF grants AST-0503871 and AST-0803158.

References

Alexander, A. L., Joner, M. D., & McNamara, D. H. (1987). *Publications of the Astronomical Society of the Pacific*, **99**, 645
Andersen, J., Clausen, J. V., & Nordstrom, B. (1984). *Astronomy and Astrophysics*, **137**, 281
Bessell, M. S. (1976). *Publications of the Astronomical Society of the Pacific*, **88**, 557
Bessell, M. S. (1979). *Publications of the Astronomical Society of the Pacific*, **91**, 589
Bond, W. C. (1850). *Annals Harvard College Observatory*, **1**, cxlix
Cousins, A. W. J. (1976). *Memoirs of the Royal Astronomical Society*, **81**, 25
Cuffey, J. (1956). *Sky and Telescope*, **15**, 258
Drilling, J. S., & Landolt, A. U. (1999). In A. N. Cox (ed.), *Allen's Astrophysical Quantities* (4th ed.), (New York: AIP Press, Springer) [Table 15.5], p. 386
Eggen, O. J. (1950a). *Astrophysical Journal*, **111**, 65
Eggen, O. J. (1950b). *Astrophysical Journal*, **111**, 81
Eggen, O. J. (1950c). *Astrophysical Journal*, **111**, 414
Graham, J. A. (1982). *Publications of the Astronomical Society of the Pacific*, **94**, 244
Hearnshaw, J. B. (1996). *The Measurement of Starlight: Two Centuries of Astronomical Photometry*. (Cambridge: Cambridge University Press)
Henden, A. A., & Kaitchuk, R. H. (1982). *Astronomical Photometry*, (New York: Van Nostrand Reinhold Company)
Iriarte, B., Johnson, H. L., Mitchell, R. I., & Wisniewski, W. K. (1965). *Sky and Telescope*, **30**, 21
Johnson, H. L. (1955). "A Photometric System," *Annales d'Astrophysique*, **18**, 292
Johnson, H. L. (1963). In K. Aa. Strand (ed.), *Basic Astronomical Data*, (Chicago: University of Chicago Press), pp. 204–224
Johnson, H. L. (1964) *Boletin de los Observatorios Tonantzintla y Tacubaya*, **3**, 305
Johnson, H. L., & Harris, D. L. III (1954). *Astrophysical Journal*, **120**, 196
Johnson, H. L., & Morgan, W. W. (1951). *Astrophysical Journal*, **114**, 522
Johnson, H. L., & Morgan, W. W. (1953). *Astrophysical Journal*, **117**, 313
Kron, G. E. (1938). *Publications of the Astronomical Society of the Pacific*, **50**, 171
Kron, G. E. (1939). *Lick Obs. Bulletin*, **19**, 53 (No. 499)
Kron, G. E. (1946). *Astrophysical Journal*, **103**, 326
Kron, G. E., & Smith, J. L. (1951). *Astrophysical Journal*, **113**, 324
Kron, G. E., White, H. S., & Gascoigne, S. C. B (1953). *Astrophysical Journal*, **118**, 502
Landolt, A. U. (1964a). *Astrophysical Journal, Supplement*, **8**, 329
Landolt, A. U. (1964b). *Astrophysical Journal, Supplement*, **8**, 352

Landolt, A. U. (1973). *Astronomical Journal*, **78**, 959
Landolt, A. U. (1983). *Astronomical Journal*, **88**, 439
Landolt, A. U. (1990). *Publications of the Astronomical Society of the Pacific*, **102**, 1382
Landolt, A. U. (1992). *Astronomical Journal*, **104**, 340
Landolt, A. U. (2007). In C. Sterken (ed.), *The Future of Photometric, Spectrophotometric, and Polarimetric Standardization*, ASP Conference Series, **364**, p. 27
Landolt, A. U. (2009). *Astronomical Journal*, **137**, 4186
Moffett, T. J., & Barnes, T. G. III (1979a). *Publications of the Astronomical Society of the Pacific*, **91**, 180
Moffett, T. J., & Barnes, T. G. III (1979b). *Astronomical Journal*, **84**, 627
Pickering, E. C. (1890). *Annals of the Harvard College Observatory*, **18**, 119
Stebbins, J., & Brown, F. C. (1907). *Astrophysical Journal*, **26**, 326
Stebbins, J., Whitford, A. E., & Johnson, H. L. (1950). *Astrophysical Journal*, **112**, 469
Sterken, C., & Manfroid, J. (1992). *Astronomical Photometry: a Guide*, (Dordrecht: Kluwer Academic Publishers)
Weaver, H. F. (1946a). *Popular Astronomy*, **54**, 211
Weaver, H. F. (1946b). *Popular Astronomy*, **54**, 287
Weaver, H. F. (1946c). *Popular Astronomy*, **54**, 339
Weaver, H. F. (1946d). *Popular Astronomy*, **54**, 389
Weaver, H. F. (1946e). *Popular Astronomy*, **54**, 451
Weaver, H. F. (1946f). *Popular Astronomy*, **54**, 504
Weaver, H. F. (1962). In *Handbuch der Astrophysik*, Band **V**, 130; *Handbuch der Physik*, Band **LIV**, 130 (1962)
Weistrop, D. (1975). *Publications of the Astronomical Society of the Pacific*, **87**, 367
Whitford, A. E. (1977). *Oral History Interview with A. E. Whitford by D. H. DeVorkin*, Amer. Inst. Phys., Niels Bohr Library, p. 35

The Rise and Improvement of Infrared Photometry

E.F. Milone and Andrew T. Young

1 Historical Developments

Infrared astronomy began with William Herschel's detection of invisible solar rays (Herschel 1800). This was followed by the development of the first pyrheliometers by Pouillet (1838); initially these instruments were thermometers covered by blackened absorbent material. A compensating instrument, in which the measurements were balanced against the heating provided by a known source, was developed by Knut Ångström (1893), who observed from Tenerife in the Canary Islands.

As reported by the Royal Observatory at Edinburgh (MNRAS, 17, 107), in 1856, Charles Piazzi Smyth detected infrared radiation from the Moon with a "thermomultiplier" (i.e., a thermopile) at Tenerife. Monitoring of the thermal infrared radiation from the Moon over its phase cycle was carried out by the Earl of Rosse in 1880. The early history of lunar thermal measurements was reviewed by Langley (1889).

After spurious detections by Huggins (1869), followed by those of several other workers, thermal radiation from Vega and Arcturus was reliably detected at the Yerkes Observatory about 1900 (Nichols, (1901)). The history of stellar infrared work has been recently reviewed by Rieke (2009).

In the early 1930s, John Hall (1934) showed that photometry near 0.87 μm could be done with a refrigerated cesium-silver-oxide gas-diode photocell. Subsequently, similar work was done by Stebbins and Whitford (1943) as reported in their series of papers on six-color photometry at Mt. Wilson. Their longest-wavelength passband had an effective wavelength of 1.03 μm. Somewhat later, Kron and Smith (1951) introduced a near-IR R, I system, whose longest passband was near 0.825 μm, again using (in modern terms) an S-1 photocathode. The availability of S-1 photocathodes enabled other astronomers to use filters that extended as far as 1 μm. Tifft (1961), for example, included a Heimann 205 filter along with a Farnsworth (16 PM-1) photocell, for the longest-wavelength passband of his 8-filter system.

E.F. Milone (✉)
RAO, Department of Physics and Astronomy, University of Calgary, 2500 University Drive, NW, Calgary, AB, Canada T2N 1N4
e-mail: milone@ucalgary.ca

E.F. Milone and C. Sterken (eds.), *Astronomical Photometry: Past, Present, and Future*, Astrophysics and Space Science Library 373, DOI 10.1007/978-1-4419-8050-2_6,

Harold Johnson (1965a, 1965b) developed passbands he called *JKL* as an extension of the *UBV* system, which Johnson and Morgan (1953) had invented to extend and improve on the accuracy of the photoelectric *P,V* system that was intended to reproduce the International System (m_{pg} and m_{pv}) of photographic photometry. Johnson's detector was a lead-sulfide cell, which has a long time-constant and low quantum efficiency. Because of the latter property, Johnson was probably not greatly concerned with tailoring the filter to exclude the more atmospherically opaque regions of the spectrum. By about 1967, the larger *JKLMNQ* suite was being used by Johnson et al. (1968), and was expanded to include the *H* passband centered around 1.65 μm, where the absorption due to the H^- ion has a minimum in the spectral energy distributions of solar-type stars. From then on, Johnson's infrared system was used primarily at national and other well-supported observatories at high-elevation sites, for reasons that we will describe in the next section.

It should be noticed that not every photometry suite that has the Johnson labels is in the infrared. For example, in 1960 Borgman (1960) introduced his *RQPNMLK* system; these passbands had effective wavelengths ranging from 0.3295 nm for *R* to 0.5880 nm for *K*. Borgman's system, therefore, was a visible-light (optical) system. Thus, not only has there been a proliferation of Johnson-type infrared passbands, but – thanks to the appropriation of some of Borgman's designations – the confusion in passband nomenclature extends beyond the infrared.

As Johnson's broad suite of passbands was coming into use, Argue (1967) and Wing (1967) each used a refrigerated RCA 7102 PMT to extend their systems into the infrared. Wing's system was a 27-passband system obtained with the Wampler scanner at the prime focus of the 36-in. Crossley Reflector of the Lick Observatory. Later, Wing used interference filters to observe in an 8-passband subset of the 27-passband system, designed to observe molecular features in stars of late spectral type (see Wing's paper in this volume). A general-purpose infrared system with three narrower passbands was used by Bahng (1969). The three were labeled *j*, *h*, and *k*, centered at 1.21, 1.59, and 2.15 μm with half-width of 0.077, 0.086, and 0.098 μm, respectively. These and many other filter systems using infrared passbands are listed in the Asiago Database on Photometric Systems (ADPS) of Moro and Munari (2000) and its on-line version: http://ulisse.pd.astro.it/Astro/ADPS/.

The development of systematic infrared photometry thus began with Johnson (1964, 1965a, 1966), but had limited use until Koornneef (1983a) examined and compiled a large volume of data from several groups, and presented a summary of intrinsic colours and calibrations for the *JHKLM* passbands (Koornneef (1983b)). Koornneef's work was extended first by Glass (1985), and then by Bessell and Brett (1888). These works attempted to obtain intrinsic near-infrared colors with well-determined transformation coefficients and zero points, and made a wider community of astronomers aware of the potential of infrared photometry (Bessell (2009)).

However, as more and more observers began to use the system, some noticed that the reproducibility and transformability were not what one might expect, given the very high precision in IR measurements that is sometimes observed over short intervals of time. [One important exception is the work by Glass and Carter (1989), who were able to correlate variations in different IR passbands (see Wing's paper in this volume for a discussion of why this is expected), and to obtain moderate transformation precision between two very closely matched sets of IR passbands.] Often when data from different observatories, or from the same observatory but from different nights, or sometimes even hours apart on the same night, were compared, results were not consistent, despite otherwise apparently photometric conditions. As a consequence, different observatories redesigned broad filters to better fit the windows at their own sites; unfortunately, these disparate variants of the Johnson system nearly all bore the same designations. Thus, what could have been a well-accepted system with widespread use became instead a confusing, multiply-defined system, presenting observers with nightmares as they tried to transform their data to an essentially nonexistent standard system.

2 Why Infrared Photometry Has been Limited to Dry, High-elevation Sites

As Forbes (1842) first noted, the heterochromatic extinction curve is not linear. The departure of the extinction curve from a straight line makes extrapolation to zero-air mass difficult. Thus the Bouguer extinction "line" (cf. Bouguer (1729)) is nonlinear in impaired passbands because the extinction varies strongly with wavelength within the passband, so that as the air mass increases, more and more of the absorption within the passband is saturated. But, as the remaining light is mostly in regions of weak absorption, the slope continually decreases more and more gradually. The situation is described by Young (1989), who discusses at length why the behaviour of the atmospheric extinction in the broad Johnson infrared passbands is so different from that in the optical passbands. Basically, this is because extinction in the infrared windows beyond about 0.8 μm is mostly saturated molecular line absorption, whereas visible extinction is mostly weak, continuous Rayleigh and aerosol scattering.

As we will show, it is possible to improve IR photometric accuracy and precision to exceed that of photometry in the visible, by carefully selecting passbands that are better fitted within the windows of the Earth's atmosphere. We will refer to the spectral regions of relative transparency as the Z, J, H, K, L, M, N, and Q windows, the near-infrared portion of which is illustrated in Fig. 1. Indeed, in the past 40 years of infrared filter use, as described in §1, placement near or within these windows is effectively the only way in which filters bearing these designations resemble Johnson's *JHKLMNQ* filters.

Manduca and Bell (1979) graphed and tabulated simulated extinction curves for Kitt Peak that demonstrated the Forbes effect. Both of the present authors independently used various mathematical functions to represent the extra-atmospheric magnitudes for these synthetic curves (Milone (1989a); Young (1989)). Young's representation involved a rational approximation. To achieve precision comparable

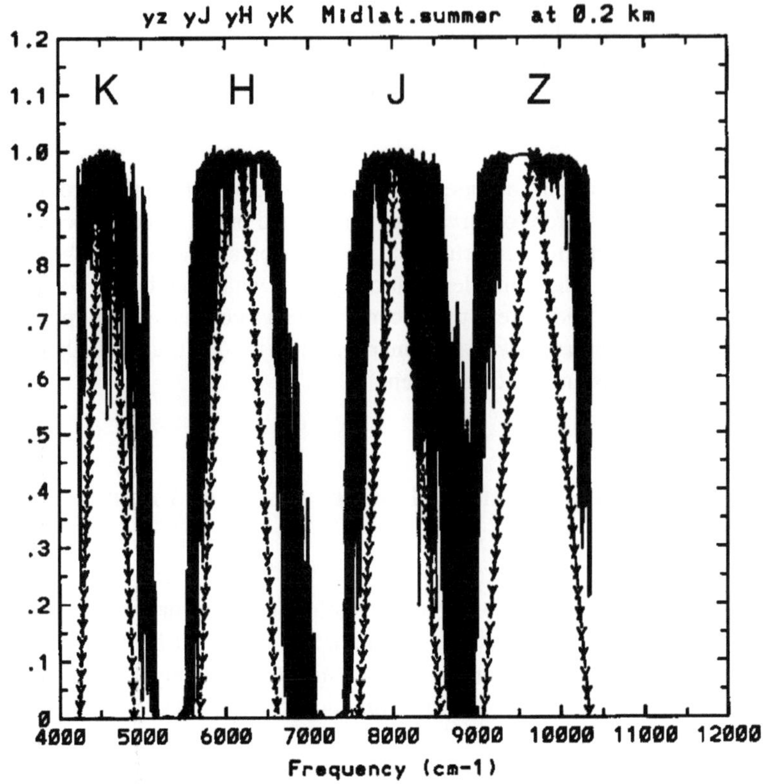

Fig. 1 The IR spectral windows of the Earth's atmosphere, in a plot of transparency vs. wavenumber. The corresponding wavelength in μm can be found by dividing 10,000 by the wavenumber expressed in cm^{-1}. The windows have been designated Z for the shortest wavelength window, J, H, and K for the others. The profiles of the passbands recommended by the Infrared Working Group of IAU Commission 25 are also shown. The adopted atmospheric model for this plot is for a mid-latitude site, in summer, and located at 200 m above sea level, approximately the elevation of the Dominion Astrophysical observatory near Victoria, B.C.

to what is routinely achieved in the visible, however, one required an "assumed" value for M_0 in the equation

$$\Delta m = \frac{(AM^2 + BM + C)}{(M + M_0)} \tag{1}$$

over the range of air mass $M = 0$ to $M = 3$. In this formulation, the extra-atmospheric magnitude is the value at $M = 0$, and the uncertainty in Δm at $M = 0$ is the quantity to be minimized. (Note the traditional use of M for airmass; some use X for air mass, and M is also used for absolute magnitude.)

Milone (1989b) contains the papers and recommendations presented at the IAU General Assembly in Baltimore in 1988 in a Joint Commission Meeting (*JCM*) involving IAU Comms. 9 (Instrumentation) and 25 (Photometry and Polarimetry). The conclusion of that meeting was that the Johnson passbands were inadequate for precise infrared photometry and needed to be redesigned to be more precisely centered within the windows of the Earth's atmosphere. It was also suggested that observers attempt to measure the water-vapor content of the atmosphere to correct for residual effects of water vapor on the photometry.

This was the origin of the collaboration later known as the "Infrared Working Group" (IRWG), beginning with the present authors, and adding members as the project expanded to include the entire community of concerned IR astronomers. The design of new passband filters accomplished the first of these tasks so well that the second became less critical to achieve. It remains an important task, however, for those who continue to use passbands that are impaired, e.g., for the study of cosmologically red-shifted spectral features.

3 The Infrared Working Group Passbands

3.1 Organization

The IRWG was formally established at the 1991 IAU General Assembly by Commission 25 president Ian McLean. One of its tasks was to evaluate all existing passbands to determine whether any had already been optimized as prescribed by the Baltimore JCM. We adopted, as a figure of merit for passbands, the rotation of a Hilbert-space vector that represents the detected spectral irradiance as it traverses the Earth's atmosphere. We called this quantity θ; the greater the change in the spectral irradiance distribution, the larger the value of θ. The physical justification for this mathematical construct is given in Young et al. (1994).

The most-impaired passbands show large changes of θ with air mass, increasing both the Forbes effect and the extinction coefficients between 1 and 3 air masses. Such impaired passbands also show strong curvature in the extinction curve between 1 and 0 air masses. We illustrate this effect for the original Johnson passbands *J* and *N* in Figs. 2 and 3, respectively. These plots were produced in the numerical experiments and simulations that are described in Sect. 3.2.

3.2 The Numerical Experiments

As spectral-irradiance sources, we adopted the full suite of Kurucz's (1989, 1993) stellar models. For terrestrial atmospheric models, we used `MODTRAN 3.7` (Berk

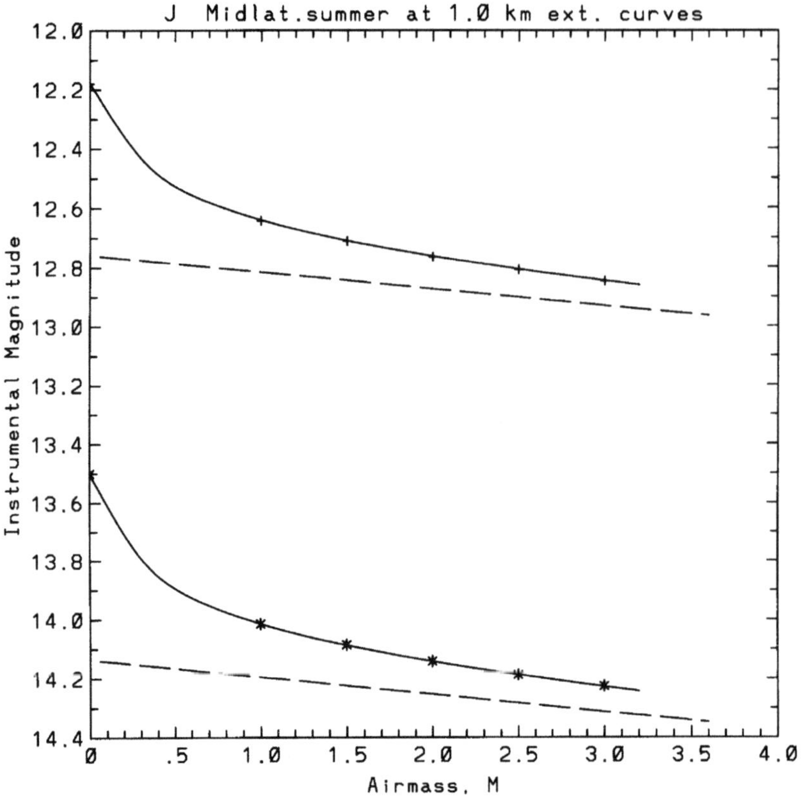

Fig. 2 Synthetic extinction curves computed for the original Johnson *J* passband. The sources for this and the other extinction plots are two of Kurucz's (1989, 1993) stellar-atmosphere models: a solar-type star (upper curve) and a red giant (lower curve); their large-air-mass asymptotes are the dashed lines. The MODTRAN atmosphere used for this and the remaining plots is the mid-latitude, summer model for a site 1 km above sea level. Note the sharp upward swing to the curves between 0 and 1 airmass, which makes a consistent linear extrapolation to an extra-atmospheric magnitude from observed data impossible

et al. (1989)). For least-squares fitting, we used the University of Texas damped least-squares code `GaussFit`. The equation of condition adopted for the extinction curves was modified from equation 1 to a slightly more convenient form:

$$m = \frac{(a + bM + cM^2)}{(1 + dM)} \tag{2}$$

In this formulation, a represents the extra-atmospheric magnitude, and $d = 1/M_0$, where M_0 is the airmass in (1) at thesharp corner of the extinction curve between

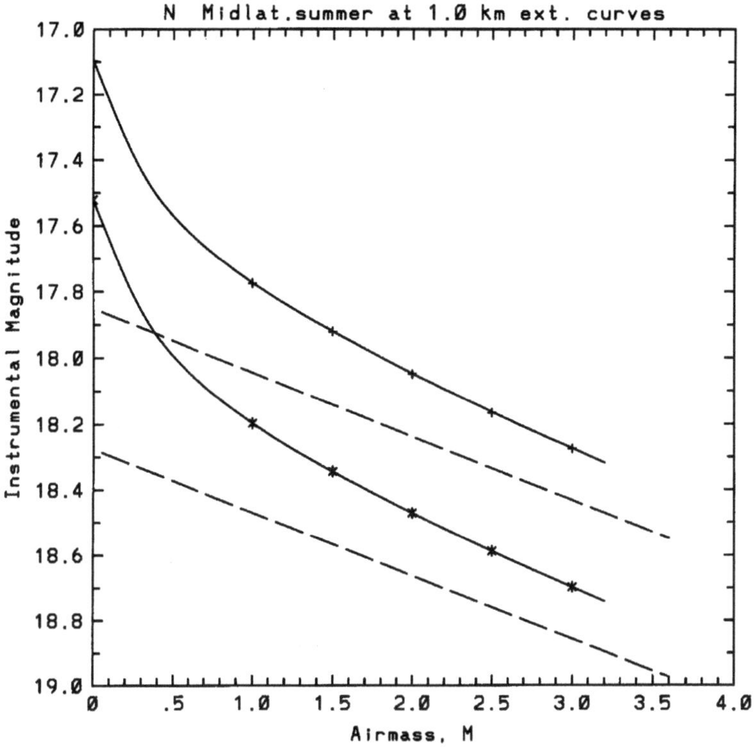

Fig. 3 Synthetic extinction curves computed for the original Johnson *N* passband in the same atmosphere used in Fig. 2. The sharp upward swing of the curves between 0 and 1 airmass makes a consistent linear extrapolation of observed data to an extra-atmospheric magnitude impossible with such an impaired passband

$M = 0$ and $M = 1$. The line

$$m = \frac{b}{d} + \frac{c}{d} \cdot M \tag{3}$$

is the large-M asymptote, which shows the slope as $M \to \infty$. The slope at $M = 0$ is $b - ad$.

Young et al. (1994) decided to further parameterize the equation, to separate coefficients that depend on source spectral distribution and/or on the atmospheric transmission from those that do not. The extinction curve representation became:

$$m = \frac{a + (a + b'_0 + b'_1 C)M + c'(d_0 + d_1 C)M^2}{1 + (d_0 + d_1 C)M} \tag{4}$$

where a is still the extra-atmospheric magnitude, the spectral- and atmosphere-dependent term $b' = b'_0 + b'_1 C = (b/d) - a$, and C is a color index. The coefficients c and d are similarly divided into a spectral- and atmosphere-dependent term and one independent of those quantities. (These are analogous to the division of k into k' and k'' terms in a linear extinction formula. However, because of the low correlation of the irregularly-distributed water-vapor lines with wavelength, such color terms offer little benefit in the IR, unless the passbands are badly centered in the atmospheric windows.)

With the stellar- and Earth-atmosphere models and code, and a few shell scripts, we ran a series of simulations for the response curves of a large number of filters in active use around the world. The coefficients of equation 4 were determined over many trials for a large sample of new and old passbands. With these, we calculated synthetic extinction curves, and found θ at 1, 2, and 3 air masses so that the filters could be compared. Finding none that fulfilled the requirements set forth at the Baltimore JCM, we then experimented with triangular and trapezoidal passband shapes of different widths and central wavelengths, and produced curves of θ vs. $\Delta\lambda$ and θ vs. λ to find the optimum widths and placements of improved passbands. The details were published by Young et al. (1994).

Milone and Young (2005) made additional comparisons with existing infrared passbands, and used the atmospheric emission within each passband to compute signal-to-noise ratios. These were further discussed in Milone and Young (2007). Most importantly, Milone and Young (2005) used filters made to the IRWG specifications to demonstrate the low extinction coefficients at a low-elevation (1,272 m) site, even in summer: less than 0.05 mag/airmass in the Z window, about 0.07 in J and K, and barely 0.03 in H.

Because the results of the numerical experiments are extensively discussed elsewhere, we mention only a few of the results here. We show the effects for a model atmosphere suited to many university and private observatories: a mid-summer, mid-latitude site at an elevation of 1 km above sea level.

3.2.1 *J* Window Passbands

The atmospheric J window for a 1-km elevation, mid-latitude summer atmospheric model, and the profiles of a few of the passbands we studied that fall near it are shown in Fig. 4. The synthetic extinction curve for the original Johnson *J* passband was shown in Fig. 2. The *rJ* and *sJ* are newer versions of the Johnson filter, better fit to the J window, but still defined to a large extent by the sides of the window. The *iJ* passband is the IRWG recommended passband for this window, and *cyJ* is the filter manufactured for this passband by Custom Scientific, Inc., of Phoenix, Arizona. The IRWG passbands were designated *yJ* and *cyJ* in Young et al. (1994), but subsequently (in Milone and Young (2005)) we have referred to them as *iJ* and *ciJ*, respectively. Extinction curves for these four passbands are shown in Figs. 5–8, respectively.

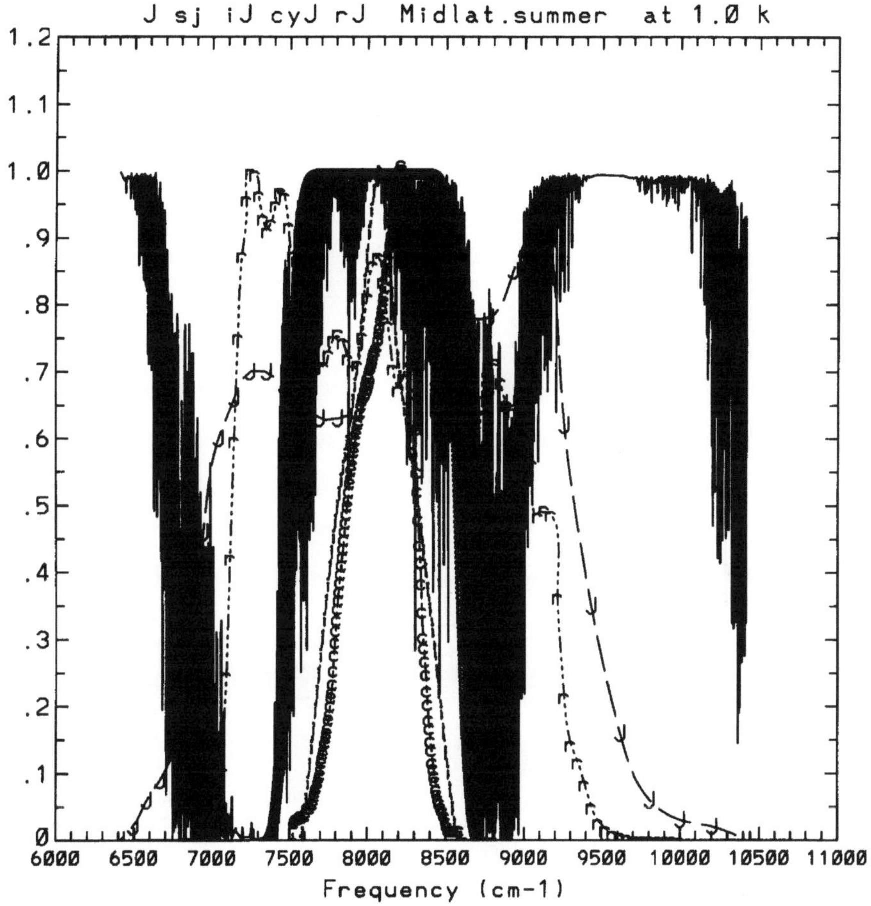

Fig. 4 The J spectral window of the Earth's atmosphere, and the profiles of a few of the passbands tested and developed by the Infrared Working Group. The MODTRAN model used for this simulation is for a mid-latitude, 1-km-elevation site in summer

3.2.2 *N* Window Passbands

The atmospheric N window for a 1-km elevation, mid-latitude summer atmospheric model, and the profiles of a few of the passbands we studied that fall near it, are shown in Fig. 9. The extinction curve for Johnson's original *N* passband can be seen in Fig. 3. A newer passband, *coN*, is in use at an observatory on Mauna Kea; the IRWG recommended passband is *yN* = *iN*. The extinction curves for these two passbands are shown in Figs. 10 and 11, respectively. Other passbands shown in the N-window plot tested the utility of a trapezoidal rather than triangular shape (*tyN*) – with very similar results to that for *yN* – and a passband in a smaller window, more impaired in the 1-km-elevation model than in models of higher sites (*yn* = *in*). Another passband in current use on Mauna Kea is *tn*.

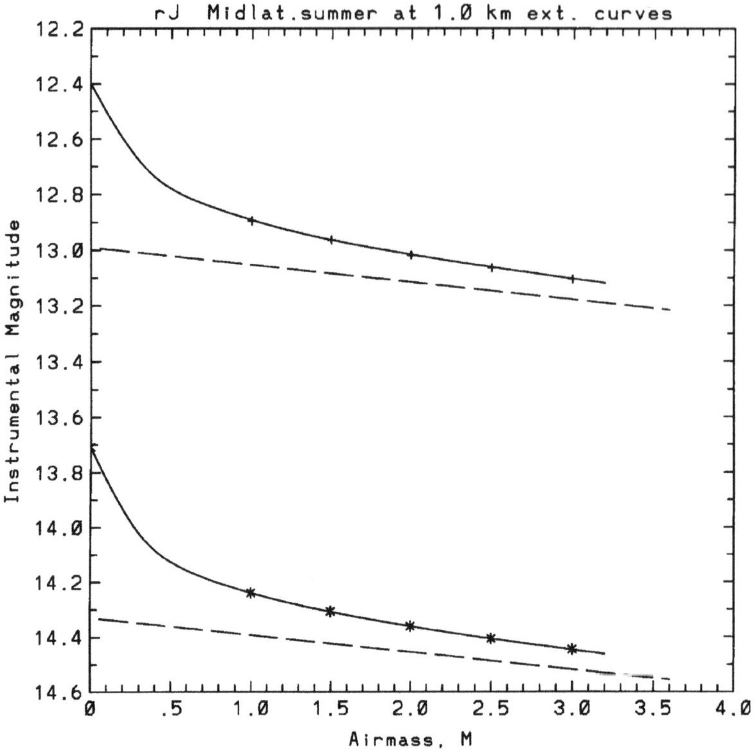

Fig. 5 Synthetic extinction curves for the *rJ* passband (a newer, somewhat narrower version of the Johnson *J* passband, but certainly not optimized to avoid water vapor absorption in any serious way). Note the strong Forbes effect; cf. Fig. 2

4 Discussion

By now most infrared astronomers have been disabused of the misconception that the Johnson-labeled infrared passbands are not greatly affected by water vapor, yet we find occasional comments in the current literature that indicate otherwise. A recent journal article included a statement that the J, H, and K_s bands are "virtually unaffected by water vapor." As we have shown in Sects. 3.2.1 and 3.2.2, Johnson's *JHKLMNQ* passbands produce large Forbes effects. These effects vary with water-vapor content on timescales of hours as well as days, and therefore are not removable unless the water-vapor content is monitored. Therefore it is pointless to reduce data to 1 air mass, as some practitioners have suggested – a practice left over from the days of low-quality visual (Pickering (1882)) and photographic photometry that persisted into the early 1950s (King (1952)), but was abandoned (Weaver (1952)) as good photoelectric data became common. In any case, it is simply wrong to state that unoptimized passbands are "virtually unaffected by water vapor."

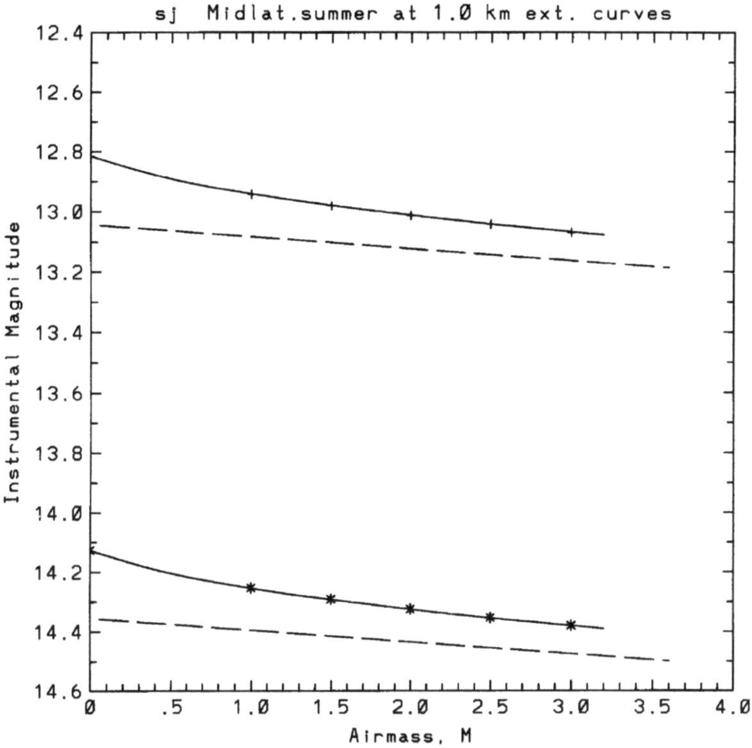

Fig. 6 Synthetic extinction curves for the *sJ* passband (a newer, narrower version of the Johnson *J* passband, but not fully optimized to avoid water-vapor absorption). Note the Forbes effect, which is less prominent than for the *rJ* passband extinction curve shown in Fig. 5

It is true that there has been an evolution away from the photometrically unsuitable original *JKLMNQ* passbands of Johnson. It is also true that, as the passbands approach the optimized IRWG passbands in placement and width, they also approach the improved photometry that this wavelength region is capable of achieving. It is *not* true that the Mauna Kea suite of near-IR passbands, or any others with which we are familiar, accomplish these goals at most photometric sites. Minimizing water-vapor absorption still leaves aerosol scattering as the main remaining source of variable extinction. However, the typical $1/\lambda$ wavelength dependence of aerosol opacity makes it less serious than in the visible region. Rayleigh scattering is $\propto \lambda^{-4}$ and so is less than 0.01 mag/airmass, even in the J window.

The low extinction coefficients and very small Forbes effects of the IRWG passbands allow observers, even at low-elevation sites under wet conditions, to use linear extinction coefficients for the *iz* and *iH* passbands; and, at 2-km and higher-elevation sites, many of the others. Differential photometry will still be possible at lower elevations, even with the 10 μm N-window passband, *iN*. Linear extinction coefficients

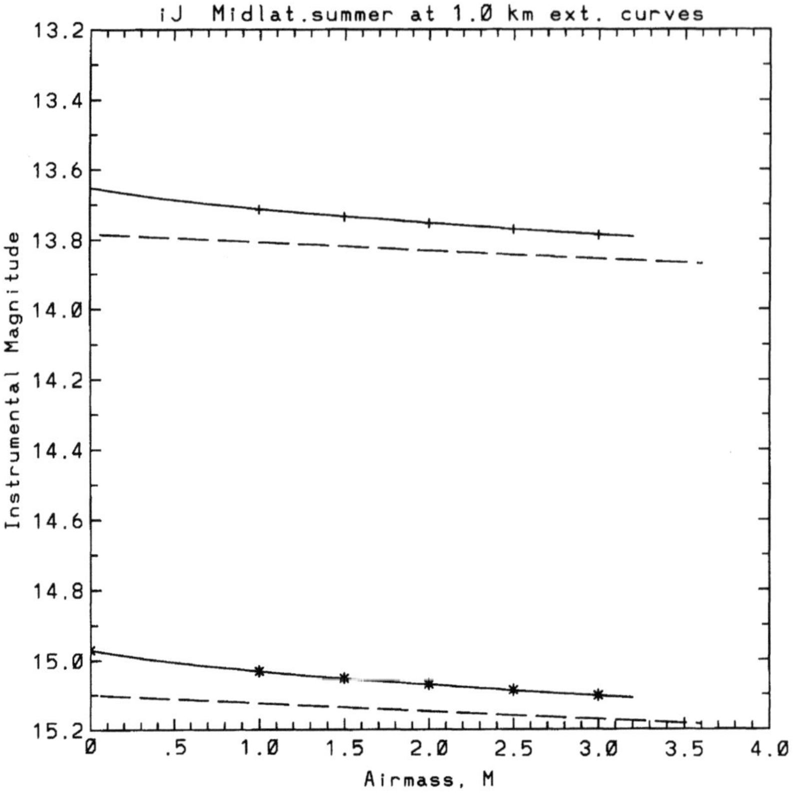

Fig. 7 Synthetic extinction curves for the *iJ* passband profile, designed to fit optimally in the *J* window. Note the very small Forbes effect

were found for the *iz*, *iJ*, *iH*, and *iK* passbands over several years at the Rothney Astrophysical Observatory, a site at 1.3 km elevation, and used to obtain systemic magnitudes and colors in these passbands (Milone and Young (2005)).

There is a price to be paid for improved photometry, and that is the main reason the IRWG filters are not in wide use at observatories around the world that currently do IR work. The edges of the IRWG passbands are not defined by the edges of the atmospheric windows: therefore, they admit no flux from these (constantly varying) edges. However, we have shown (Milone and Young (2005, 2007)) that a measure of the signal-to-noise ratio varies inversely with extinction and with a measure of the Forbes effect. So, the small loss of raw throughput is recouped in signal-to-noise gain. This is the point that at least some IR facility operators have failed to grasp, thus far. We suggest that if the IRWG passbands are made available, they will be used!

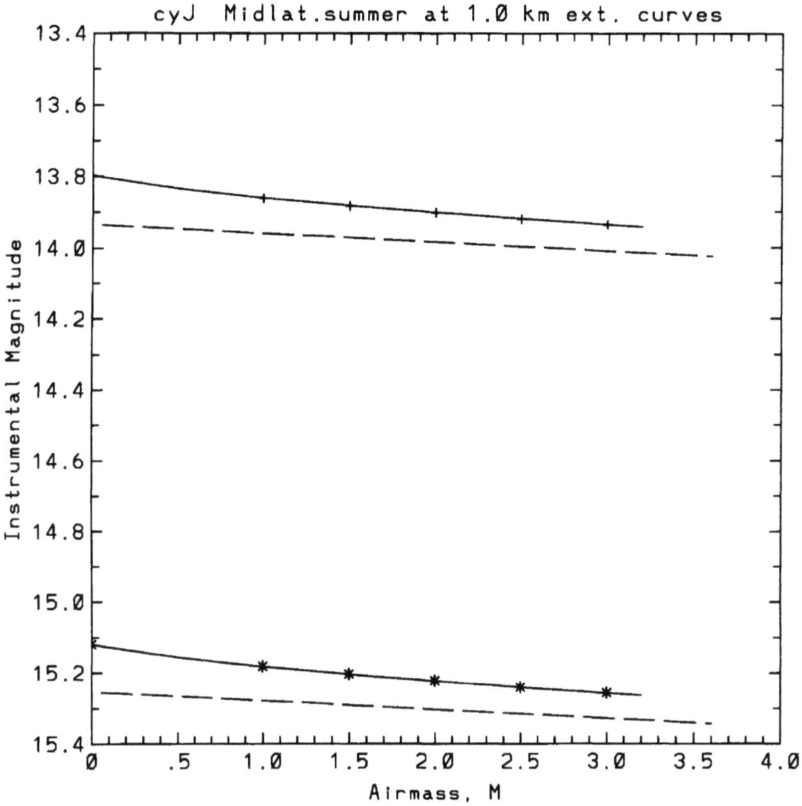

Fig. 8 Synthetic extinction curves for the *ciJ* = *cyJ* passband, using a filter manufactured by Custom Scientific to the specifications of the *iJ* passband (cf. Fig. 7). Note the very small Forbes effect

5 Conclusion

Infrared photometry is important for revealing the universe beyond what visual detectors capture. It can provide even higher precision and accuracy than visual photometry, but only if the effects of varying water-vapor absorption in the infrared passbands are minimized. This has been the goal of the Infrared Working Group of IAU Commission 25; the means to do so have been found and reported (Young et al. (1994)), as mandated by the IAU Joint Commission Meeting in 1988. In Milone and Young (2005, 2007, 2008) we demonstrated that the IRWG passband system provided lower extinction values, decreased Forbes effects, higher signal-to-noise ratios, and better figures of merit than almost all previous passbands in use that bear the Johnson designations. We also presented observational data that show the practicality of the near-infrared bands of the IRWG system.

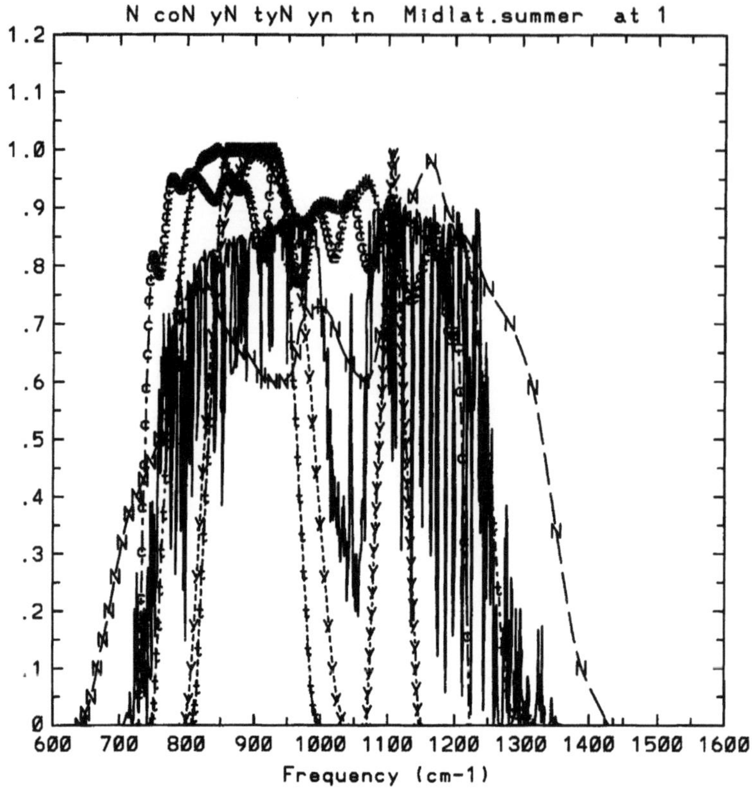

Fig. 9 The N spectral window of the Earth's atmosphere, and the profiles of some passbands studied by the Infrared Working Group. We show the original Johnson *N*; a later version of this passband, *coN*; a much newer version used at Mauna Kea; and the synthetic optimized passband *iN* (coded here as "yN"). The "*in*" (coded here as "*yn*") is a narrower IRWG passband centered near 1,100 cm^{-1}. The same MODTRAN model used in Figs. 2–11 is shown

Finally, although all these works demonstrate the potential for improved photometric accuracy and precision at all sites, in the last two papers we argue that the greatest potential gain of the IRWG system is for sites which have not previously undertaken infrared photometry, but where precise photometry is currently possible at visible wavelengths. Thanks to sharply decreased Forbes effects at these sites, differential photometry may be carried out with ease, and even absolute photometry may be performed, if conditions are photometric. It should be noted that the manufacturer of the IRWG filters offers discounts on bulk orders, which lowers the premium that had to be paid in the past to use these passbands.

It should be borne in mind that although the IRWG passbands facilitate photometry of higher precision and accuracy than any other passbands intended for broadband ground-based work have done, they cannot guarantee good photometry.

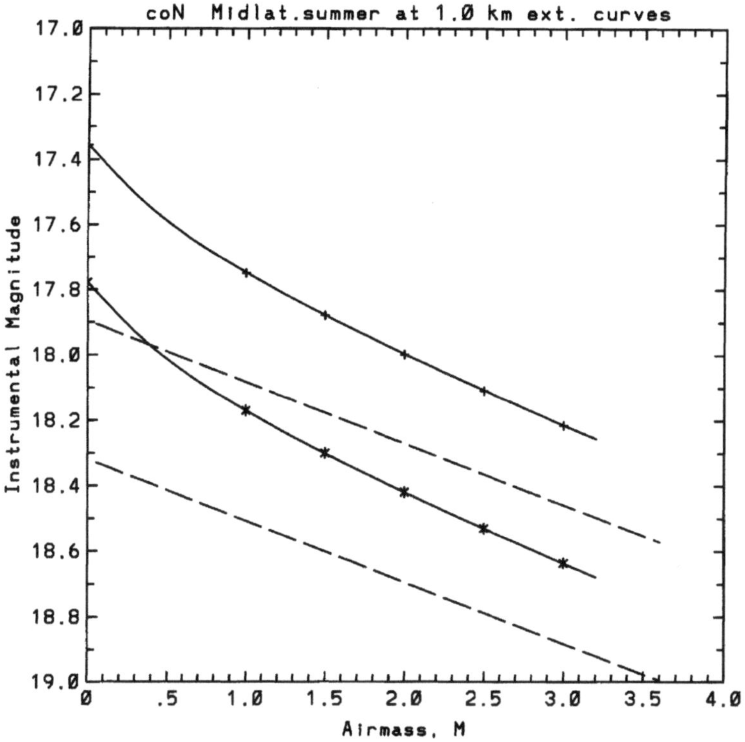

Fig. 10 Synthetic extinction curves for the *coN* passband profile, a newer version of the Johnson *N* passband designed to fit more precisely in the N window as observed at Mauna Kea. Note the smaller Forbes effect than for the Johnson *N* passband, shown in Fig. 3

Photometry in these passbands, as in all others, is limited by auroral emissions, aerosols, and by clouds, and, in any case, still requires careful and systematic observing practices.

Acknowledgements The support of the Physics & Astronomy Department of the University of Calgary has been critical to the success of the IRWG experiments and results. It is a pleasure to thank undergraduate astrophysics majors in the senior astrophysics laboratory course that EFM taught for more than a decade, especially the summer assistants who were recruited from that course, among whom were: Kyle Degenhardt, Kerry Dubray, Danielle Fraser, Christopher Ostrowski, Jon Ramsey, Russell Shanahan, Caleb Sundstrom, and Christopher Winder, for conscientious assistance with observations and data reduction. Dr. Philip Langill, then Resident Astronomer at the RAO, and Dr. Alfredo Louro also assisted with data acquisition. Over the past decade, Michael Williams improved telescope software and helped train observers. RAO technician, Frederick Babott maintained all the RAO instrument systems, including the infrared, since their inception; most recently, that work was taken over by James Pike. Much of the work described here was supported by grants to EFM from the Natural Sciences and Engineering Research Council and the National Research Council of Canada, which, as always, are gratefully acknowledged. This paper is No. 78 in the Publications of the RAO reprint series.

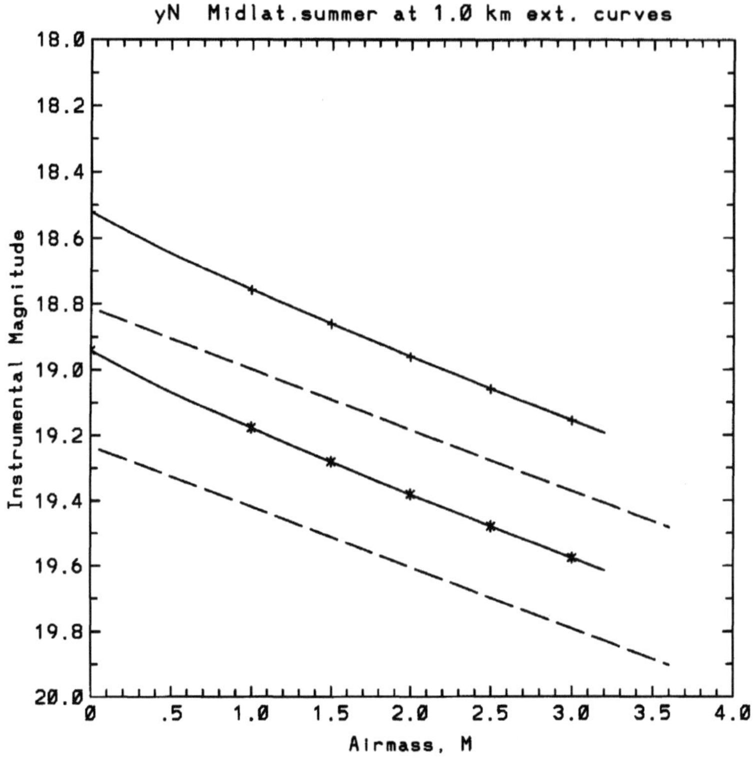

Fig. 11 Synthetic extinction curves for the $yN = iN$ passband profile, optimized for the N window. Note the still smaller predicted Forbes effect than in Fig. 10. The linearity of the curve permits differential photometry of high precision, with careful color-matching of target and comparison stars. A small difference in the Forbes effect for the two different stellar models can be seen

References

Ångström, K. (1893). *Physical Review*, **1**, 365

Argue, A. N. (1967). *Monthly Notices of the Royal Astronomical Society*, **135**, 23

Bahng, J. (1969). *Monthly Notices of the Royal Astronomical Society*, **143**, 73

Berk, A., Bernstein, L. S., & Robertson, D. C. (1989). *MODTRAN: A Moderate Resolution Model for LOWTRAN 7* (GL-TR-890122), (Bedford, MA: Air Force Geophysics Laboratory)

Bessell, M. S. (2009). *Astronomy and Astrophysics*, **500**, 257

Bessell, M. S., & Brett, J. R. (1888). *Publications of the Astronomical Society of the Pacific*, **100**, 1134

Borgman, J. (1960). *Bulletin of the Astronomical Institutes of the Netherlands*, **15**, 255

Bouguer, P. (1729). *Essai d'Optique sur la Gradation de la Lumière*. (Paris: Claude Jombert)

Forbes, J. D. (1842). *Philosophical Transactions of Royal Society of London*, **132**, 225

Glass, I. S. (1985). *Irish Astronomical Journal*, **17**, 1

Glass, I. S., & Carter, B. S. (1989). "Infrared Extinction at Sutherland," in E. F. Milone (ed.), *Infrared Extinction and Standardization*, (Berlin: Springer-Verlag) *Lecture Notes in Physics*, Vol. **341**, p. 37

Hall, J. S. (1934). *Astrophysical Journal*, **79**, 145

Herschel, W. (1800). Investigation of the power of the prismatic colours to heat and illuminate objects. *Philosophical Transactions Royal Society of London*, **132**, 255
Huggins, W. (1869). *Proceedings of the Royal Society of London*, **17**, 309
Johnson, H. L. (1964). *Bol. Obs. Tonantzintla y Tacubaya*, **3**, 305
Johnson, H. L. (1965a). *Communications, Lunar and Planetary Laboratory*, **3**, 73
Johnson, H. L. (1965b). *Astrophysical Journal*, **141**, 923
Johnson, H. L. (1966). *Annual Review of Astron and Astrophys*, **4**, 193
Johnson, H. L., & Morgan, W. W. (1953). *Astrophysical Journal*, **117**, 313
Johnson, H. L., MacArthur, J. W., & Mitchell, R. I. (1968). *Astrophysical Journal*, **152**, 465
King, I. (1952). *Astronomical Journal*, **57**, 253
Koornneef, J. (1983a). *Astronomy and Astrophysics, Supplement*, **51**, 489
Koornneef, J. (1983b). *Astronomy and Astrophysics*, **128**, 84
Kron, G. E., & Smith, J. L. (1951). *Astrophysical Journal*, **113**, 324
Kurucz, R. L. (1989). "Reducing Photometry by Computing Atmospheric Transmission", in E. F. Milone (ed.), *Infrared Extinction and Standardization*, (Berlin: Springer-Verlag) *Lecture Notes in Physics*. Vol. **341**, p. 55
Kurucz, R. L. (1993). In E. F. Milone (ed.), *Light Curve Modeling of Eclipsing Binary Stars*, (New York: Springer-Verlag), p. 93
Langley, S. P. (1889). *Memoirs of National Academy Science*, **4**, Part 2
Manduca, A., & Bell, R. A. (1979). *Publications of the Astronomical Society of the Pacific*, **91**, 848
Milone, E. F. (1989a). "Problems of Infrared Extinction and Standardization — An Introduction," in E. F. Milone (ed.), *Infrared Extinction and Standardization*, (Berlin: Springer-Verlag) *Lecture Notes in Physics*, Vol. **341**, p. 1
Milone, E. F. (ed.), (1989b). *Infrared Extinction Standardization*, (Berlin: Springer-Verlag) *Lecture Notes in Physics*, Vol. **341**, p. 1
Milone, E. F., & Young, A. T. (2005). *Publications of the Astronomical Society of the Pacific*, **117**, 485
Milone, E. F., & Young, A. T. (2007). In C. Sterken (ed.), *The Future of Photometric, Spectrophotometric, and Polarimetric Standardization*, (ASP Conf. Series) Vol. **364**, p. 387
Milone, E. F., & Young, A. T. (2008). *Journal of the American Association Variable Star Observers*, **36**(1), 110
Moro, D., & Munari, U. (2000). *Astronomy and Astrophysics, Supplement*, **147**, 361
Nichols, E. F. (1901). *Astrophysical Journal*, **13**, 101
Pickering, E. C. (1882). *Astronomische Nachrichten*, **102**, 193
Pouillet, C.-S. (1838). *Comptes Rendus de l'Acad. des Sciences*, **7**, 24
Rieke, G. H. (2009). *Experimental Astronomy*, **25**, 125
Stebbins, J., & Whitford, A. E. (1943). *Astrophysical Journal*, **98**, 20
Tifft, W. G. (1961). *Astronomical Journal*, **66**, 390
Weaver, H. F. (1952). *Astrophysical Journal*, **116**, 612
Wing, R. F. (1967). "Infrared Scanner Measurements of the Colors and Molecular Band Strengths of Late-Type Stars," in M. Hack (ed.), *Proceedings of the Colloquium, on Late-Type Stars*, (Trieste: Osservatorio Astronomico), p. 231
Young, A. T. (1989). "Extinction and Transformation," in E. F. Milone (ed.), *Infrared Extinction and Standardization*, (Berlin: Springer-Verlag) *Lecture Notes in Physics*, Vol. **341**, p. 6
Young, A. T., Milone, E. F., & Stagg, C. R. (1994). *Astronomy and Astrophysics, Supplement*, **105**, 259

On the Use of Photometry in Spectral Classification

Robert F. Wing

1 Introduction

The techniques of photometry have had roles to play in nearly all aspects of stellar astronomy, and have nearly always resulted in important improvements in accuracy. It is therefore not unreasonable to ask what role photometry can play in the classification of stellar spectra, and what accuracy can be achieved by this means.

It is, however, important to understand that photometry and spectroscopy, as traditionally practiced, are fundamentally different techniques, both in method and in objective. We will begin in Sect. 2 with general considerations of the differences between photometry and spectroscopy, and the distinction between "photometric information" and "spectroscopic information." In Sects. 3 and 4 we look at several examples of the use of wide-band and intermediate-band filter photometry to obtain "spectroscopic information," i.e., information that is intrinsic to the star. Finally, the writer's eight-color narrow-band system of TiO/CN classification photometry is discussed in Sect. 5 as an example of a photometric system that provides both photometric and spectroscopic information.

2 General Considerations

Classification spectroscopy and multicolor photometry are both techniques that are commonly used to study individual stars, often with small telescopes and modest equipment. Both techniques fall within the province of IAU Commission 45 on Stellar Classification (originally known as the commission on "Spectral Classification and Multi-band Colour Indices") – but there the similarity ends!

R.F. Wing (✉)
Department of Astronomy, Ohio State University, Columbus, OH, USA
e-mail: wing@astronomy.ohio-state.edu

E.F. Milone and C. Sterken (eds.), *Astronomical Photometry: Past, Present, and Future*, Astrophysics and Space Science Library 373, DOI 10.1007/978-1-4419-8050-2_7,

2.1 Spectroscopy vs. Photometry

Spectral classification is traditionally done by comparing the spectra of program stars to those of standard stars, observed with the same spectrograph; this comparison is normally done by visual inspection, and the results are qualitative in nature. Photometry, on the other hand, is a quantitative technique, providing a set of numerical magnitudes in various filters. Because photometric magnitudes and colors (magnitude differences) are routinely measured with great precision, it would seem reasonable to ask whether the use of photometry can improve the accuracy of spectral classification. However, the precision of photometry is not necessarily even relevant to the objectives of spectral classification.

Spectroscopic and photometric observers have different objectives, different concerns at the telescope, and, some would say, even different personalities. The spectroscopist, needing to inspect small details of the spectrum such as the relative strengths of weak spectral lines, is very concerned with spectral resolution and instrumental focus but is oblivious to the thin clouds that may be floating past the dome – they don't affect the resolution or the equivalent widths of spectral lines. The photometrist, needing to compare a star's brightness to those of standard stars scattered over the sky, requires stable equipment and a near-perfect sky but can tolerate imperfect focus – after all, resolution is hardly relevant to the measurement of integrated quantities such as the light transmitted by a filter.

It is useful to make a distinction between the types of information obtained from spectroscopic and photometric observations. What I call "spectroscopic quantities" are those obtained from the spectrum that are intrinsic to the star. Classification on the MK system provides two pieces of spectroscopic information, namely a spectral type and a luminosity class (and sometimes notes about other characteristics). A classification such as K2 III tells us roughly the star's temperature and luminosity; it tells us nothing about the star's distance or about any intervening interstellar material. "Photometric quantities," on the other hand, are magnitudes and colors; magnitudes are obviously affected by distance and by the absorption caused by any interstellar dust that lies in the line of sight, and colors are affected by the wavelength dependence of the interstellar absorption. When both types of information are available, the reddening can be determined by comparing the observed color to the color normally associated with the star's spectral type; the observed magnitude can then be corrected for interstellar extinction (by assuming a normal reddening law); and finally the distance can be determined by comparing the corrected apparent magnitude to the absolute magnitude associated with the star's luminosity class. This is the method of "spectroscopic parallax." It depends on having both "spectroscopic" and "photometric" information about the star – i.e., data both intrinsic to the star and extrinsic to it.

A classic example of the use of both spectroscopic and photometric information to derive distances was the study of 1270 O- and B-type stars by Morgan and colleagues at Yerkes Observatory (Morgan et al. 1953, 1955). They combined MK classification with two-color photoelectric photometry to compute reddening values and distances for these young stars, thereby providing the first direct evidence of the spiral structure of the Milky Way galaxy. We will see below that the writer's eight-color system of TiO/CN classification photometry provides both photometric and spectroscopic information for M stars, so that it also can yield reddening-corrected distances, in this case for M-type supergiants (MacConnell et al. 1992).

2.2 Can Photometry Give a "Spectral" Classification?

Many photometric systems can be used to classify stars, i.e., to sort them in various ways, and to recognize various kinds of unusual stars. Some of these systems are quite sophisticated, being able to circumvent the effects of interstellar reddening and to produce indices that are strongly correlated with MK spectral types and/or luminosity classes. Still, classifications that are based upon photometric indices and their correlation with spectral types are "photometric" classifications, which are not the same as true *spectral* classifications. But what is the difference?

Photometric classifications can be very precise, sorting stars into finely-graded divisions. In favorable cases, photometric classifications and their calibration have resulted in some of the best determinations of temperature and luminosity available. The accuracy that can be achieved in such determinations has been discussed by many authors – notably by Strömgren (1963), Golay (1974), Crawford (1975), and Straižys (1992) – and will not be discussed again here. Rather, I will give examples from several photometric systems to show how intrinsic information about stars can be obtained from multicolor photometry, and to illustrate the differences between photometric and spectral classification.

What constitutes a "true spectral classification?" Under what circumstances, if any, can a classification based on photometry be considered a true spectral classification? When a classification is published, can it legitimately be given in the MK notation if it is based on something other than the "MK process?" Does the reader need to be told how the classification was obtained? These questions often arise when stellar spectroscopists and photometrists get together.

The answers to such questions remain elusive, largely because the borderlines between spectroscopy and photometry can be indistinct. It is easy to say that a spectral classification must be based on an observation of the spectrum. But if photometrists observe a star with a set of several filters, are they not observing the spectrum? I believe the distinction must be based on the type of spectral feature that is employed in the classification. Spectral classifications are based on the observation of *specific* spectral features. These are usually atomic lines, or line ratios – features that are usually much too small to be measured by filter photometry – but they also include molecular bands, which can be measured by carefully-placed narrow-band filters.

I would propose the following criteria for the circumstances under which a classification based on photometry can properly be labeled a "spectral classification":

- The photometry must provide "spectroscopic information" – quantities that are intrinsic to the star, i.e., independent of reddening or distance.

- The classification criteria should be specific spectral features that are sensitive to physical characteristics – usually temperature and/or luminosity – or to some aspect of the chemical composition.
- Even when these conditions are met, a photometric index will be a valid classification criterion only within a limited range of temperature and luminosity (i.e., in a restricted region of the HR diagram), or for only a certain kind of star.

Many photometric systems meet the first criterion and provide "spectroscopic information," and several examples will be given in what follows. The second criterion, however, is harder to meet. Filters that are wide enough for efficient observing nearly always include a large number of spectral lines rather than a specific spectral feature. Classifications that are based on photometric indices of the slope of the spectrum (or a change in slope), or on the position of the star in some diagram, should be considered "photometric types."

Unfortunately, these criteria have not always been followed. There are cases in the published literature where spectral types, in the MK notation, have been reported that are based solely on wide-band photometry. For example, there is a well-established correlation between the $B-V$ color and spectral type, and authors who have measured $B-V$ may be tempted to report an equivalent spectral type based on this correlation. Unfortunately, such equivalent types are not always clearly identified as photometric types, and they may be picked up by compilers of catalogues and mixed with legitimate spectral classifications. One problem is that the observed $B-V$ color is not an intrinsic property of the star, as it is sensitive to any interstellar reddening that may be present. Another problem concerns the limited range of applicability: the correlation between $B-V$ and spectral type applies strictly only to main-sequence stars of a given metallicity, and if the star has some other luminosity or metallicity, the use of the correlation will yield an incorrect spectral type, even if the reddening is zero (or known). The spectroscopist, in assigning a classification, has an opportunity to check the star's luminosity class and approximate metallicity and can take these into account. The photometrist, knowing only the $B-V$ color, has none of these advantages. Clearly, it is misleading to use the term "spectral classification" if the only observed quantity is a photometric color.

The above example is an almost trivial case of the possible confusion between photometric and spectral classifications. We should, however, keep this distinction in mind as we consider the information content of various photometric systems.

3 Examples from Wide-Band Photometry

A great many photometric systems have been used to sort out stars in various ways, to estimate values for various physical parameters, and to look for candidates for certain unusual kinds of stars. This is "classification" of stars in the general sense, although not strictly "spectral classification." Extensive reviews of the material have been compiled by, among others, Strömgren (1963), Golay (1974), and Straižys (1992). In most cases, the filters have not been placed to measure specific spectral features.

In this Section and the next, I will discuss a few examples to show how intrinsic properties of stars can be obtained from wide-band and intermediate-bandwidth photometry. We will then move on to the narrow-band system of TiO/CN photometry with which I have been involved for many years, the filters of which *have* been placed to measure specific spectral features.

3.1 Comments on the Use of *UBV* Photometry

Observations on a photometric system of N filters are usually presented in the form of one magnitude and $N-1$ color indices – that is, one distance-dependent quantity and $N-1$ quantities that are independent of distance. Such a system provides the possibility of determining at most $N-1$ intrinsic properties of the star. Of course, different color indices may give redundant information, and often do. For example, if all filters are placed at continuum points (i.e., if they are little affected by spectral features), then all color indices will give the same information (the color temperature of the star), and observations plotted in any color–color diagram will fall along a straight line, the slope of which depends only upon the spacing of the filters.

When Harold Johnson developed the *UBV* system (Johnson and Morgan 1953), he was providing a color index ($B-V$) that would serve the same function as the traditional color index $CI = m_{\rm pg} - m_{\rm vis}$ formed by comparing photographic and visual magnitudes, but with the improvement that ultraviolet light was excluded from the blue filter. He also realized that the addition of a U filter, with an effective wavelength of approximately 3,600 Å, would add significantly to the information content of the system – i.e., that he was not merely adding a redundant color index.

The information content of the *UBV* system is shown by its color–color diagram (Fig. 1). Both colors increase for redder stars, but note that the $U-B$ color increases downwards. Filled circles represent the intrinsic colors of main-sequence stars from Johnson (1966), who lists them as a function of spectral type. The decidedly non-linear shape of the intrinsic relation tells us at once that the $U-B$ color contains different information from that of the $B-V$ color.

The arrow in Fig. 1 shows the slope of the reddening line, given by the ratio of the excesses in the two color indices due to interstellar reddening:

$$E(U-B)/E(B-V) = 0.72. \tag{1}$$

This is the slope along which any star will be moved in the diagram if there is (normal) interstellar dust along the line of sight.[1] Also shown in Fig. 1 is a (dashed) line, with a slightly steeper slope than the reddening line, representing the locus of

[1] For simplification here, I am taking this ratio to be a constant, as it would be if the effective wavelengths of the filters were constant; a more exact treatment (Johnson 1958) allows for the fact that the effective wavelengths of all three filters depend on the color of the light source, resulting in a reddening line with some curvature.

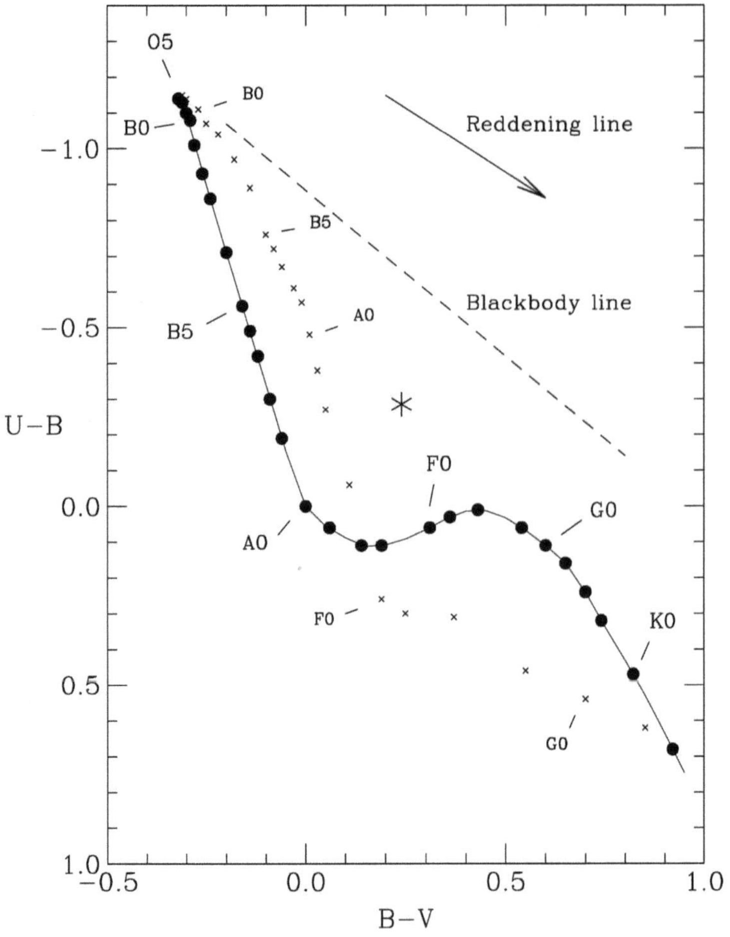

Fig. 1 The color–color diagram of *UBV* photometry. The plotted points are the intrinsic colors of main-sequence stars (*filled circles*) and supergiants (*crosses*) as determined by Johnson (1966) as a function of spectral type. Also shown are the slopes of the *reddening line* and the *blackbody line*. The star (*asterisk*) at the center of the figure is a fictitious object discussed in the text

blackbodies of different temperatures. This line, too, is actually somewhat curved in an exact treatment (for the same reason), but the straight-line approximation is good enough to show roughly where stars would fall if they radiated like blackbodies in the *UBV* filters.

For early-type stars, the difference between the blackbody line and the intrinsic relation is almost entirely the result of absorption in the *U* filter, and presumably this is why it has always seemed natural to plot this diagram with the $U-B$ color increasing downwards. This absorption is caused by neutral hydrogen in the $n = 2$ level, which absorbs both in the higher lines of the Balmer series as they converge to a limit at 3,646 Å, and in the Balmer continuum beyond that limit. From the hottest

O–type stars, in which H is almost completely ionized, to the stars of type A, the absorption in U increases steadily as H becomes increasingly neutral. After type A, H is completely neutral so that ionization no longer plays a role, and absorption from the $n = 2$ level decreases as the temperature drops. The stellar relation then moves back toward the blackbody line but does not reach it because of the onset of strong absorption by lines of neutral metallic atoms. Of course, metallic lines also affect the B filter; in stars of types F, G, and K, both $U-B$ and $B-V$ are reddened by metallic lines, displacing stars downward and to the right in the diagram (Wildey et al. 1962).

If a normal main-sequence star of type O or B is reddened, its colors will be shifted along a line of slope 0.72 in the color–color diagram; the star can be "de-reddened" by sliding it back along a line of this slope until it intersects the intrinsic relation. One can then "read off" the star's intrinsic colors and hence spectral type.

The "Q–method" of UBV photometry, introduced by Johnson and Morgan (1953), is the algebraic equivalent of the graphical process described above. They define the quantity Q as

$$Q = (U - B) - 0.72\,(B - V) \tag{2}$$

where the coefficient 0.72 is the slope of the reddening line. Because the color excess caused by any amount of reddening will be 0.72 times as great on $U-B$ as on $B-V$, the effect of reddening on Q will be zero.[2] For stars belonging to the straight-line portion of the color–color plot (note that this straight line passes through the origin because the $U-B$ and $B-V$ colors are both zeroed to stars of type A0 V), the Q parameter bears a simple relation to the intrinsic $(U - B)_\circ$ color:

$$Q = 0.80(U - B)_\circ \tag{3}$$

It then follows that the intrinsic color $(U - B)_\circ$ can be expressed as a function of the observed colors:

$$(U - B)_\circ = 1.25(U - B) - 0.90(B - V). \tag{4}$$

Then, given the intrinsic $U-B$ color, one can look up the corresponding spectral type. The result, however, is a "photometric" type. It should not be called a "spectral" classification because the spectrum has not been observed. Even if the star is known to be a main-sequence star, the possibility of, say, a chemically-peculiar star cannot be ruled out by the photometry alone.

For main-sequence stars, the Q–method (either graphical or algebraic) produces spectral types of approximately the same accuracy as those derived from direct MK classification. The problem, of course, is that the photometry alone cannot test the hypothesis that the star is a normal main-sequence star. The method is most useful when applied to star clusters, because the color-magnitude diagram of a cluster can

---

[2] Analogous reddening-independent quantities can be formed for any three-color photometric system; to determine the appropriate value of the coefficient, one needs only the filter wavelengths and an interstellar reddening law (Johnson 1968).

be used to identify stars belonging to the cluster main sequence. The study of the open cluster M29 by Morgan and Harris (1956) illustrates the application of the Q–method to a cluster, both graphically and algebraically.

The intrinsic colors of supergiants determined by Johnson (1966) are plotted as small crosses in Fig. 1. The values are difficult to determine and somewhat uncertain because all supergiants are distant and likely to be reddened; however, there is no doubt that the intrinsic relation for early-type supergiants lies between the main-sequence relation and the blackbody line. One's first impression is that the supergiants are shifted to the right in Fig. 1, but upon identifying the spectral types corresponding to each plotted point, we see that early-type supergiants are shifted vertically with respect to main-sequence stars of the same spectral type. Supergiants are brighter in U because their Balmer lines are much narrower, allowing light to escape between the lines.

Finally, to illustrate the problems of classifying stars by wide-band photometry alone, let us consider the object plotted as a large star near the center of Fig. 1 at $B-V=0.25, U-B=-0.30$. We have its observed colors, but no other information. What can we say about it? It could be a main-squence star of spectral type B5 V, reddened by $E(B-V)=0.40$. But it could equally well be a supergiant of type A0 or A1, with only half as much reddening. A classification spectrogram would have no trouble distinguishing between these possibilities, but the colors alone cannot. And those are not the only possibilities. If the star lies at a high galactic latitude, it is unlikely to be reddened, but then it could be a distant quasar, or a nearby white dwarf – anything with less absorption in the ultraviolet than normal stars have. Also, it could be a metal-poor F-type dwarf, displaced from the normal relation by the weakness of its metallic lines.

UBV photometry has had its greatest successes in the study of galactic clusters that are young enough to contain O- and B-type stars, for then the Q–method can separate the effects of temperature and reddening. The method works basically because of a combination of factors: (*a*) the optical spectra of early-type stars are very clean, with only one spectral feature – absorption by neutral H – that is strong enough to affect colors measured with wide-band filters; (*b*) this feature is measured by the U filter, giving the *UBV* system a measurement of a feature intrinsic to the star; and (*c*) the strength of this feature increases steadily with spectral type from O5 to A0, causing the intrinsic relation to have a very different slope from that of the reddening line. But for cooler stars the Q–method breaks down, as the U filter alone cannot distinguish between decreasing absorption from the $n=2$ level of hydrogen and increasing absorption from atomic lines, and as the intrinsic relation runs approximately parallel to the reddening line. To separate the effects of temperature, luminosity, metallicity, and reddening in cooler stars where all these parameters affect the spectrum, we need more than three filters, and they must be narrower and carefully placed in the spectrum.

3.2 *Infrared JHK Photometry*

When Johnson introduced the wide-band filters *J* (1.25 µm), *H* (1.65 µm), and *K* (2.2 µm), his intention was to extend the *UBVRI* system into the infrared for the purpose of measuring stellar energy distributions. Filters were chosen to (roughly) match the windows in telluric absorption, without regard to stellar spectral features. Because of difficulties in acquiring a suitable filter for the window around 1.65 µm, the *H* filter was the last of these to be added and consequently is out of alphabetical order – see Wing (1994) for a review of these historical developments.

As it turned out, the *JHK* filters provide much more than just three points on stellar energy distributions. Here we consider what these filter offer as a means of classifying stars according to intrinsic stellar properties.

The *JHK* filters are normally thought of as a unit, because they employ the same detector and are conveniently observed and reduced together. Indeed, many studies have been based on *JHK* photometry alone. As an extreme example, the *Two Micron All Sky Survey* (2MASS – Cutri et al. 2003) has provided *JHK* photometry for several hundred *million* stars. What can we learn from it?

As with any three-color photometric system, we can examine the information content by plotting the color–color diagram, in this case $J-H$ vs. $H-K$. If the plot is limited to giants and supergiants, the points fall roughly along a diagonal line, telling us that the two color indices are strongly correlated. Thus to a large extent, both indices are providing the same information, namely the color temperature of the star. And because the slope of the giant relation is not very different from that of the reddening line, we may conclude that *JHK* photometry will not be very useful for separating the effects of temperature and reddening. However, from observations of giant stars in globular clusters of known metallicity and reddening, Frogel et al. (1983) have shown that the width of the observed relation is attributable to metallicity differences.

When dwarf stars are added to the *JHK* color–color diagram, an interesting feature appears. Beyond about spectral type M0, the sequence for M dwarfs diverges sharply from that of M giants and supergiants (Bessell and Brett 1988). Thus M dwarfs can be securely identified by *JHK* photometry alone.

In an effort to explain the bifurcation of the dwarf and giant sequences in the *JHK* color–color diagram, synthetic spectrum calculations have been carried out at the Niels Bohr Institute (Jørgensen and Wing 2000; Wing and Jørgensen 2003). We considered the separate effects of several molecules that have bands in the near-infrared – CO, CN, metallic oxides, and H_2O – but found their effects to be much too small to account for the splitting of the sequences. Rather, the splitting has more to do with the effect of H^- opacity and the different temperature structures of dwarf and giant atmospheres.

Regardless of the interpretation, it is empirically obvious that *JHK* photometry is a useful classification device, especially when applied to large data sets such as the 2MASS catalogue. But can *spectral* classifications be given on the basis of *JHK* photometry? If a star has the *JHK* colors of, say, an M4 dwarf, there is almost no possibility that it is anything but an M4 dwarf. Still, without an observation of the spectrum or the measurement of any specific spectral feature, I would argue that this is a photometric classification, not to be mixed with true spectral classifications.

Optimized versions of these filters have been designed and tested by the Infrared Working Group of IAU Commission 25 (Milone and Young 2005, and in the present volume). By excluding regions of significant (and variable) extinction due to atmospheric water vapor, these filters allow observations from sites at lower elevations and yield reductions of greater precision. However, the basic information content of three-color photometry in the J, H, K windows is not expected to change through use of the optimized filters, because that is governed largely by the effects of stellar H^- opacity.

4 Intermediate-Band Photometry

We have seen that UBV photometry is most successful in determining intrinsic stellar properties for stars earlier than type A0, but encounters difficulty with cooler stars. This is the consequence of two changes that occur after about type A0:

1. Absorption in the U band by neutral H passes through a maximum at type A0 and then weakens rapidly. Consequently the Q–method no longer works to separate the effects of temperature and reddening (the quantity Q is still independent of reddening, but it is no longer uniquely related to the intrinsic $(U-B)_\circ$ color).
2. Absorption by neutral metallic lines grows steadily after type A, affecting B more than V, and U more than B. Increased metallic-line absorption therefore reddens both $U-B$ and $B-V$, making its effects nearly impossible to untangle from those of interstellar reddening.

The problem is basically that a three-color system can determine at most two intrinsic properties, whereas the colors of, say, normal solar-type stars are affected by three parameters – temperature, reddening, and metallicity – even when the luminosity is known, as in cluster work.

Several intermediate-bandwidth photometric systems, with bands of about 200–400 Å in width and four or more in number, have been used to attack this problem. Whether a four-color system can actually determine three independent quantities will, of course, depend on how the filters are placed in the spectrum. Here I will comment on just a few of these systems.

4.1 Strömgren uvby Photometry

The four-color *uvby* (ultraviolet, violet, blue, yellow) system designed by Strömgren (1963, 1966) has been widely used and is particularly successful in determining metallicities of F- and G-type stars. Let's consider how it works.

Metallic lines are relatively weak in the visual region and have little affect on the y filter, but they increase in density to shorter wavelengths. This increase, however, is not linear; indeed, if line absorption increased to shorter wavelengths at a constant rate, its effect would mimic that of interstellar reddening and the two effects would not be separable, regardless of the number and placement of the filters. Fortunately, the density of atomic lines changes rather suddenly at about 4,500 Å, producing a change in the slope of the spectrum that is easily seen on low-dispersion spectrograms. The B filter of the UBV system straddles the region where this change occurs, but Strömgren realized that if he divided the blue region into two parts, with v and b filters on either side of 4,500 Å, he could measure the change of slope and use it as an index of metallicity. This is the function of the m_1 index, defined as

$$m_1 = (v-b) - (b-y). \tag{5}$$

Here the $(v-b)$ color, with filters at 4,110 and 4,670 Å, is sensitive to metallicity as well as to temperature, whereas the $(b-y)$ color, involving filters entirely longward of 4,500 Å, is sensitive only to temperature.

Of course, interstellar reddening affects both terms of (5), by unequal amounts. Actually, the increase is larger on the second term (due to its wider spread in wavelength), so that reddening makes the m_1 index smaller. A reddening-free version of the m_1 index, denoted $[m_1]$, is defined in a manner analogous to the Q–method of UBV photometry:

$$[m_1] = m_1 + 0.18(b-y). \tag{6}$$

In a similar manner, the u filter at 3,500 Å can be combined with other filters to form the c_1 index measuring thc change in slope around 4,000 Å (which in B stars is caused by neutral hydrogen absorption):

$$c_1 = (u-v) - (v-b), \tag{7}$$

and this index also has a reddening-free version:

$$[c_1] = c_1 - 0.20(b-y). \tag{8}$$

The $[m_1] - [c_1]$ diagram is the basic classification device of the Strömgren four-color photometry (see Fig. 1 of Strömgren 1966). The relation for main-squence stars is well-defined but highly non-linear. Regions of the diagram can be mapped out in terms of two-dimensional MK classes, although the spectral types assigned on this basis are, of course, photometric types.

An interesting application of the *uvby* photometry makes use of the $c_1 - (b-y)$ diagram (see Fig. 4 of Strömgren 1963). For stars of types A and F, the c_1 index is sensitive to luminosity and $b-y$ serves as a temperature index, so that this diagram represents a section of the HR diagram. Strömgren's observations of 1,100 bright stars of types A2–G0 show a well-defined lower envelope which can be identified with the zero-age main sequence; stars lying above it are interpreted as having started their evolution off the main sequence. In this way, small departures from the zero-age main sequence can be recorded much more precisely than by conventional spectroscopic classification.

Observations on the *ubvy* system are often accompanied by narrow-band measurements of the hydrogen line Hβ. Crawford (1958) showed that a reddening-free Hβ index could be obtained accurately and efficiently by using two filters, one much wider than the other, both centered on the line. The Hβ index is a sensitive luminosity discriminant, particularly in stars of type F. Numerous open clusters and associations have been studied by *ubvy*β photometry, especially by Crawford at Kitt Peak. For calibrations of the system and discussion of the accuracy achieved, see the series of papers by Crawford (1975, 1978, 1979).

4.2 The Vilnius and Geneva Systems

At the Vilnius Observatory, Straižys (1963) developed an eight-color, medium-bandwidth system for the purpose of classifying stars of all types. Careful consideration was given to the placement of the filters with respect to the major spectral features, although the bandwidths chosen (mostly between 200 and 300 Å) necessarily measure spectral intervals rather than individual features. In order to be useful over a wide range of temperature, the filters extend over a relatively wide range in wavelength, from approximately 3,500–6,500 Å. A summary of results from the Vilnius system has been given by Straižys (1992), who also gives a complete bibliography (see his Table 61). The system is suitable for surveys in which stars of all types may be encountered; it is also successful in recognizing various types of peculiar stars.

A parallel development at the Geneva Observatory was initiated by Golay (1963). This photometric system, which became known as the Geneva system, consists of 7 filters – the 3 wide-band filters of (essentially) the *UBV* system, and 4 filters of medium bandwidth. The latter divide the *B* and *V* filters into two parts each. Note that the division of the *B* filter into two parts gives the Geneva system the capability of determining metallicity in F, G, and K stars in the same manner as the Strömgren system. Large numbers of stars have been observed on this system by Rufener, Grenon, and others at the Geneva Observatory (see Straižys, 1992 for an extensive bibliography).

In recognition of the similarities in the filters and objectives of the Vilnius and Geneva systems, the observers of the two institutions compared notes and decided on a single set of 7 filters that has nearly the same information content as the Vilnius system but also the fainter limiting magnitude of the Geneva system. The result is called the VILGEN system (Straižys 1979).

4.3 DDO Photometry

A six-color system introduced by McClure and van den Bergh (1968) at the David Dunlap Observatory is well suited to work on relatively late-type stars, namely those of types G and K. Two of its filters, relatively narrow (80 Å) ones centered at 4,166 and 4,257 Å, provide a measure of the (0,1) band of CN, which degrades to shorter wavelengths from a head at 4,216 Å. This feature is sensitive to luminosity in G and K stars, and it is discussed further in Sect. 5.7.1 below. The DDO system also has filters at 4,517 and 4,886 Å which are used to form a "gravity index" (McClure 1970). However, because the 4,517 Å filter coincides with the (0,2) band of CN, it should not be considered surprising that the DDO gravity index is strongly correlated with the DDO CN index. This system has been used primarily for studies of open clusters, and population studies in the Galaxy (McClure 1979, and references therein).

4.4 Cambridge Narrow-Band Photometry

An important series of papers from Cambridge Observatory reported narrow-band measurements of a number of specific spectral features. A grating spectrometer, wavelength mask, and photomultipliers were used to define photometric indices of strong atomic lines and molecular bands. The first paper of the series (Griffin and Redman 1960) describes the instrumentation and reports measurements of the (0,1) band of the Violet System of CN, to which we will return in Sect. 5.7.1. In terms of the criteria proposed in Sect. 2.2, the Cambridge measurements of specific spectral features would be suitable for use in deriving true spectral classifications, but they haven't been employed in that manner. Rather, the Cambridge observers mostly used wide-band $R-I$ colors as their temperature indicator (the objects of study being nearby unreddened dwarfs and giants) and judged luminosities from published MK classifications; their narrow-band indices of spectral features were then used to determine abundance parameters.

5 Narrow-Band TiO/CN Classification Photometry

There exists one particular system of narrow-band photometry that has been developed specifically for the purpose of providing spectral classifications. This is the eight-color (8c) system of TiO/CN classification photometry introduced by the author (Wing 1971) and used for many years with photomultiplier tubes serving as the detector; the system is still in use, but now with CCD detectors (Wing 2005). The first observations on this system were made in 1969, and it was the recognition of the system's 40-year anniversary (Wing 2009) that led to the inclusion of this discussion in the present book.

In light of comments made in Sect. 2 of this chapter, I should start by explaining why I consider the classifications based on this system to be true spectral classifications, as opposed to "photometric classifications" which rely upon a correlation between color and spectral type. We will then examine the extent to which the eight-color photometric system can reproduce the MK types given for giant stars by P. C. Keenan. Finally, I will discuss the accuracy of the system and the limitations on its applicability.

5.1 Objectives of the Eight-Color System

In developing the eight-color system, my objective was to obtain both photometric and spectroscopic information for stars of type M, whose optical spectra are dominated by bands of the TiO molecule. As discussed earlier, by "photometric information" I mean magnitudes and colors, measured preferably in regions as free as possible from molecular blanketing. "Spectroscopic information," on the other hand, refers to the measurement of specific spectral features.

The eight-color system uses reddening-free indices of the strength of TiO and CN absorption bands as indicators of temperature and luminosity, respectively. By using these spectroscopic quantities to give a two-dimensonal classification, we obtain a true *spectral* classification and can then employ the separate "photometric information" of the system to derive the reddening and distance.

One might think that a photometric system designed to deal with stars of a single spectral type (M) should be thought of as a "special-purpose" system of very limited applicability, even after it is found that stars of the later K subtypes can also be classified. We should remember, however, that the K and M dwarfs are the most numerous stars in the solar neighborhood, and probably in any volume-limited sample. It is also true that in magnitude-limited samples defined at relatively long wavelengths, such as the *Two-Micron Sky Survey* (Neugebauer and Leighton 1969), the most numerous stars are the M-type giants and supergiants. Furthermore, the brightest stars of globular clusters are K (and sometimes M) giants, while the brightest stars of younger populations (open clusters) are often giants or supergiants of type M. Clearly there are a great many possible applications for a photometric system that produces spectral classifications for M stars, especially in view of the difficulties encountered by MK classification when the molecular absorptions are strong.

I think it is only by remarkably good luck that nature has provided us with a pair of molecules – TiO and CN – whose optical and near-infrared bands are suitable for the two-dimensional (temperature/luminosity) classification of M stars. It is also fortunate that the one-micron region, which includes the flux maxima of most M stars, also contains stretches of spectrum that are relatively free from molecular absorption, as well as some of the strongest bands of these two molecules.

Table 1 Properties of filters of the eight-color system

Filter	λ_c	$\Delta\lambda$	Feature measured	Contaminants
1	7120	60	TiO $\gamma(0,0)$ band	weak CN
2	7540	50	continuum	weak CN; VO after M6
3	7810	40	TiO $\gamma(2,3)$ band	
4	8120	50	CN (3,1) band	
5	10395	50	continuum; I (104)	
6	10540	60	VO for SpT > M6	
7	10810	60	continuum	
8	10975	70	CN (0,0) band	

5.2 Definition of the System

The characteristics of the interference filters defining the eight-color system are listed in Table 1. The filters were designed with widths of 40–70 Å to fit within regions depressed by the molecules to be measured, staying away from sharp spectral discontinuities. The wavelength interval includes the clean one-micron region and its well-defined VO features which degrade longward from 10,460 Å (see Fig. 6 of Spinrad and Wing 1969); the interval extends shortward to 7,120 Å to include the strong (0,0) band of the TiO γ–system (all TiO bands of longer wavelength are weaker), and it extends longward to 10,975 Å to measure the (0,0) band of the Red System of CN. The whole system of fluxes is zeroed to Vega (α Lyr) at filter 5, and the flux at filter 5 is sometimes called the I (104) magnitude.

At least 11 sets of these filters have been produced over the years – in 1969, 1971, and a multiple order in 1976 which included sets installed at KPNO and CTIO. These were small (half-inch diameter) filters suitable for observing one star at a time with photoelectric photometers. More recently (in 2001) two sets of large-format (two-inch diameter) filters have been made for use with CCD imaging detectors; these are currently in use at CTIO and the Turkish National Observatory. Because the available CCD detectors do not reach wavelengths as long as those of filters 7 and 8, the CCD system (Wing 2005) includes only the first 6 filters. It has been found, however, that giving up the measurement of the CN (0,0) band does not affect the quality of the classifications, as filter 4 provides a satisfactory CN index.

5.3 Questions to be Addressed

In order to assess the accuracy that can be obtained in the spectral classification of M stars though use of the eight-color system of TiO/CN classification photometry, it is not sufficient (or particularly interesting) to discuss the photometric accuracy. With the system's well-iterated set of standard stars (Wing 1979; MacConnell et al. 1992) and with observations from good photometric sites, it is routine to obtain photometric accuracies of ±0.01 mag in the fluxes and still better accuracy in the colors and indices. The interesting questions that we need to address are more complicated:

- How good is TiO as a temperature indicator? To what extent is it affected by other parameters, e.g., luminosity and metallicity? Over what range in spectral type can it be usefully measured?
- Likewise for CN, how sensitive is it to luminosity, and how *in*sensitive to temperature? Over what area of the HR diagram can it be effectively used?
- How well do classifications based on TiO and CN correlate with MK classifications? Which method has the best internal precision?

There are also practical questions that must be answered:

- Given the eight-color data for a star, viz., a set of 8 magnitudes reduced to outside the earth's atmosphere through observations of standard stars, how can we define indices of TiO and CN strength that are independent of reddening and thus suitable for use as classification criteria?
- In view of the fact that both TiO and CN have extensive, overlapping band systems, are our indices of these molecules really independent of one another, as would be required for a truly orthogonal two-dimensional classification system?

5.4 MK Standard Stars

Most of these questions can best be answered by consideration of the set of 60 well-observed, bright MK standard stars that have been used to calibrate the eight-color system in terms of spectral type. These stars are listed in Table 2. In order to obtain a uniform set of calibration stars, I have restricted it to stars classified as class III giants by P. C. Keenan. I have also weeded out a few stars noted by Keenan as peculiar – e.g., weak-line stars, or stars showing enhancements of *s*–process elements. Thus the sample should represent the typical Population I giants of the solar neighborhood.

The second column of Table 2 gives Keenan's MK classification. Most of these are from Morgan and Keenan (1973) although some have been taken from Keenan (1963), Landi Dessy and Keenan (1966), and Keenan and Pitts (1980). The eight-color data for these stars, in columns 3–6, include the number N of observations, the indices of TiO and CN band strength (see below) obtained from the mean 8c fluxes, and the two-dimensional spectral classification obtained from these indices. The photometric data for these stars were obtained prior to 1980, when both KPNO and CTIO operated telescopes small enough (0.4-m) to allow observations of such bright stars. Multiply-observed non-variable stars ($N > 10$ or so in Table 2) served as photometric standards as well as classification standards.

Table 2 MK Standards Used as Calibration Stars

Star	SpT(MK)	N	TiO (mmag)	CN (mmag)	SpT(8c)	Star	SpT(MK)	N	TiO (mmag)	CN (mmag)	SpT(8c)
α Ari	K2− IIIab	28	0	91	< K3.5	83 UMa	M2 IIIab	7	460	78	M1.7 III
ε CrB	K2 IIIab	25	0	91	< K3.5	χ Peg	M2+ III	59	494	86	M2.0 III
ρ Boo	K3 III	30	4	103	< K3.5	74 Vir	M2+ III	6	530	77	M2.1 III
α Hya	K3 IIIa	16	38	119	K3.5 II	8 And	M2+ III	3	557	83	M2.3 III
β Cnc	K4 III	36	61	96	K4.1 III	λ Aqr	M2.5 IIIa	2	496	83	M2.0 III
β UMi	K4 III	12	71	97	K4.2 III	η^2 Dor	M2.5 III	2	706	67	M3.0 III
HR 4991	K4 III	2	91	93	K4.4 III	HR 6337	M3− III	2	628	91	M2.6 III
ρ Ser	K4.5 III	2	153	92	K5.0 III	ψ Vir	M3− III	2	643	72	M2.7 III
γ Dra	K5 III	5	147	101	K5.0 II	ψ Peg	M3 III	2	713	94	M3.0 III
α Tau	K5 III	11	218	86	K5.7 III	ρ UMa	M3 IIIb	7	717	106	M3.1 II
υ Boo	K5+ III	3	195	96	K5.5 III	HR 46	M3 III	2	742	97	M3.1 III
31 Lyn	K7 III	3	176	89	K5.3 III	μ Gem	M3 IIIab	11	777	72	M3.3 III
α Lyn	K7 IIIab	3	191	83	K5.4 III	30 Psc	M3 III	2	826	65	M3.4 III
17 Per	K7 III	3	216	73	K5.7 III	δ Vir	M3+ III	4	797	75	M3.3 III
γ Sge	M0− III	17	210	123	K5.6 II	ρ Per	M4 IIIa	6	1102	69	M4.3 III
υ Gem	M0 III	3	247	96	M0.0 III	10 LQ Her	M4 III	3	1268	78	M4.7 III
γ Eri	M0 III	17	261	75	M0.1 III	ω Vir	M4–4.5 III	7	1219	85	M4.5 III
μ UMa	M0 III	4	263	91	M0.1 III	TU CVn	M5− III	4	1421	59	M5.1 III
β And	M0 IIIa	6	296	89	M0.5 III	AC Dra	M5− IIIa	2	1326	57	M4.8 III
75 Leo	M0 IIIb	5	296	108	M0.5 II	RR UMi	M5 III	5	1234	71	M4.6 III
κ Ser	M0.5 IIIab	3	275	80	M0.2 III	HD 77443	M5 III	4	1551	65	M5.4 III
δ Oph	M0.5 III	9	327	86	M0.8 III	HD 11961	M5 III	3	1675	61	M5.7 III
36 Com	M1− IIIb	3	312	97	M0.6 III	56 VY Leo	M5.5 III	12	1752	66	M5.9 III
ν Vir	M1 IIIab	20	318	58	M0.7 III	45 RZ Ari	M6− III:	8	1772	65	M5.9 III
α Vul	M1 IIIb	2	319	82	M0.7 III	30 g Her	M6− III	5	1897	46	M6.3 III
55 Peg	M1 IIIab	37	334	100	M0.8 III	EU Del	M6 III	3	1848	57	M6.2 III
37 Leo	M1 III	5	375	97	M1.2 III	BK Vir	M7− III	9	2203	106	M7.0
ϕ Aqr	M1.5 III	3	421	112	M1.5 II	EP Aqr	M7 III:	2	2057	49	M6.5 III
α Cet	M1.5 IIIa	26	453	89	M1.7 III	RX Boo	M7.5−M8	13	2247	23	M8.0
π Leo	M2− IIIab	8	449	101	M1.7 II	R Dor	M8e III:	5	2144	98	M7.3

5.5 *Definition of Indices*

With multicolor photometry, there are several ways to define reddening-free indices. We have already discussed the *Q*–method of *UBV* photometry and the similarly-defined reddening-free $[m_1]$ and $[c_1]$ indices of the Strömgren four-color system. In the eight-color TiO/CN photometry, I prefer a different approach, one which I think is easier to visualize. A reddening-free index of the absorption in any filter can be defined by comparing the observed flux to the flux in the continuum at the same wavelength. To obtain the continuum value at the wavelength of the observed feature, we must decide on the best way to interpolate or extrapolate from the observed continuum points. Clearly this method will work only if the photometric system includes filters placed at reasonably good continuum points.

My approach has been to reduce the observations to a system of absolute fluxes. Then the reduced photometry, when plotted against wavelength, has the character of a "spectrum", albeit one of only 8 data points. The mean data for five of the calibration stars, representing a progression in spectral type, are plotted in Fig. 2. Because the data are given on an absolute flux scale, they can be compared directly to blackbody curves, or to synthetic spectra computed from model atmospheres. In particular, we can fit a blackbody curve through the fluxes at the best continuum points (filters 2 and 6 in M stars) and use this as the continuum at the other filters.

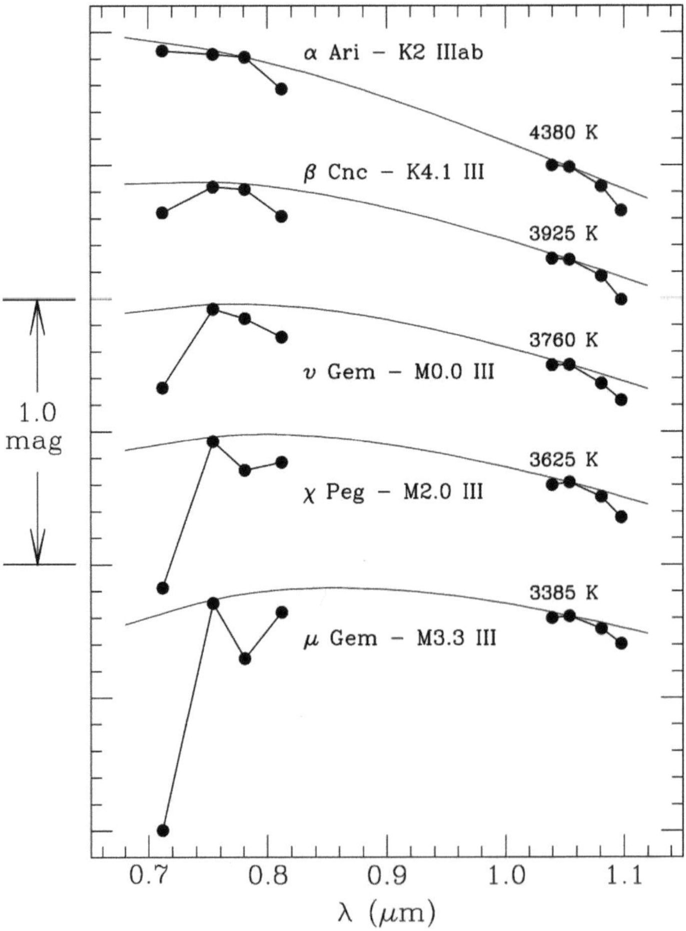

Fig. 2 Eight-color data for 5 calibration stars (each plotted with an arbitrary additive constant). Each has been fitted with a blackbody curve, the temperature of which is indicated. TiO depresses filters 1 and 3 and is present in all but the topmost star. CN depresses filters 4 and 8 and has approximately the same strength in all 5 of these giant stars

Of course, the actual continuum of an M star does not have exactly the shape of a blackbody, as it is modified by H^- opacity; but a blackbody continuum is relatively simple to compute – and certainly better than a straight line!

With the blackbody continuum in hand, simple subtraction of the observed flux from the continuum flux (in magnitudes) gives a reddening-free index of the absorption at each filter. Thus the "depression" $D(n)$ at filter n is the observed flux $F(n)$ *minus* the corresponding continuum value $C(n)$:

$$D(n) = F(n) - C(n) \tag{9}$$

(remembering that the magnitude scale increases to fainter values); then the TiO index is defined as the depression at filter 1, whereas the CN index is the mean of the depressions at filters 4 and 8:

$$\begin{aligned} TiO &= D(1) \\ CN &= 0.5[D(4) + D(8)] \end{aligned} \tag{10}$$

But there is a problem. Not surprisingly, because the band systems of TiO and CN overlap considerably, there is CN contamination in both the primary TiO filter (# 1) and, to a lesser extent, the continuum filter (# 2). We must correct for this contamination if we want to have a TiO index that is truly independent of CN. This is important, since M-type dwarfs, giants, and supergiants have systematically different CN strengths, and one of my original objectives in setting up the eight-color system was to place the spectral types of all M stars on the same TiO–based scale, regardless of luminosity.

So we carry out an iteration. The CN index defined as above is used to calculate the corrections, $\Delta_{\mathrm{CN}}(1)$ and $\Delta_{\mathrm{CN}}(2)$, to be applied to filters 1 and 2, respectively, to correct for CN contamination:

$$\begin{aligned} \Delta_{\mathrm{CN}}(1) &= 0.55\, CN \\ \Delta_{\mathrm{CN}}(2) &= 0.20\, CN \end{aligned} \tag{11}$$

There corrections, in magnitudes, are applied (by subtraction) to the fluxes in the corresponding filters:

$$F^*(n) = F(n) - \Delta_{\mathrm{CN}}(n) \quad n = 1,2 \tag{12}$$

to obtain the CN–corrected fluxes F^* in filters 1 and 2, and a second blackbody fit is made through the corrected flux in filter 2. The final TiO index is then the difference between the corrected flux in filter 1 and the flux in the new continuum. In Fig. 2, it can be seen that the blackbody curves pass slightly above the data points at filter 2: the observed fluxes are plotted, and the blackbody curves plotted and used in the analysis pass through the corrected flux $F^*(2)$. These corrections are quite small – the effect on the TiO index is zero for M dwarfs, typically about 0.03

mag in M giants, and about twice that in M supergiants – but they remove what otherwise would be a systematic difference in the classification scales. The TiO and CN indices given, in Table 2 are the corrected indices – that is, the indices computed from the second blackbody fit. (The indices, like the quantities $F(n)$, $C(n)$, and $D(n)$, are normally expressed in magnitudes to three decimals; but in Table 2 the indices have been multiplied by 1,000 so that leading zeroes and decimal points could be removed to save space in the crowded table).

5.6 Effectiveness of TiO as a Temperature Indicator

The strength of TiO absorption has always played a dominant role in the classification of M stars. In the early work at Harvard (e.g., Cannon and Pickering 1914), M stars were defined to be those showing TiO bands in objective-prism spectra, and they were subdivided into classes Ma, Mb, and Mc on the basis of TiO strength. The somewhat more refined numerical subdivisions M0 to M6 used at Mount Wilson (Adams et al. 1926) were "based almost entirely on the intensities of the bands of titanium oxide, only minor consideration being given to other features." By the time of the Mount Wilson work it was generally assumed, although not yet proved, that the TiO strength was governed by the photospheric temperature. With the introduction of the MK system, Morgan et al. (1943) adopted the Mount Wilson numerical subdivisions but, for consistency with their scheme for classifying warmer stars, they sought temperature-sensitive atomic line ratios as their definitive temperature criteria. The MK Atlas includes a plate entitled "The M Sequence Is A Temperature Sequence," on which the authors demonstrate that the TiO band strength does in fact correlate closely with their chosen temperature–sensitive line ratios.

Two decades after the introduction of the MK system, Keenan (1963) gives a more complete description of the optical spectra of M-type giants and supergiants. His temperature criteria for the subdivisions of M stars, defined in 1963 and used in all his subsequent publications, include both atomic-line ratios and TiO band intensities, although he concedes that with advancing type, the TiO strength necessarily plays an increasingly dominant role.

An important practical difficulty experienced by the classical MK observers is one imposed by the limited dynamic range of photographic plates, forcing the observer to use different TiO bands at different subtypes. By contrast, narrow-band photometry can successfully measure both very weak and very strong bands, and the 7,120 Å filter of the 8c system follows the growth of a single TiO band over an enormous range of band strength.

Calibration of the 8c TiO index in terms of spectral type has been done by plotting the index against MK type for the calibration stars of Table 2. That plot is not shown here, as it has already been presented by MacConnell et al. (1992). The tightness of this relation for Keenan standard stars shows that Keenan's types – which he tried to make independent of TiO by basing them as much as possible on atomic-line ratios – are in fact very closely correlated with TiO strength.

A close look at Fig. 2 shows that TiO absorption is clearly present at 7,120 Å in stars as early as type K4. In fact, the TiO index for β Cnc (MK type K4 III; 8c type K4.1 III) is 0.06 mag, considerably more than the error of measurement. The adopted relation between TiO index and spectral subtype is a series of straight-line segments, the parameters of which are given in MacConnell et al. (1992). The slope of the segment between K4.0 and K5.9 is 0.1 subtype per 0.01 mag in the index. After type M0.0 (which is defined to immediately follow K5.9) the slope increases, until each 0.1 subtype in the segment from M4.0 to M6.0 corresponds to a change of 0.04 mag in the index. It is therefore no exaggeration to present temperature classes to 0.1 subtype for any type after K4.0. The index itself grows from 0.05 mag at K4.0 to a quarter-magnitude at type M0.0, a half-magnitude at M2.0, a full magnitude at M4.0, and 1.8 mag at M6.0.

To address the effectiveness of the 8c TiO index as an indicator of temperature in a two-dimensional (temperature/luminosity) classification scheme, we need to consider: (1) how well the 8c temperature classes reproduce the MK types of class III giants classified by P. C. Keenan; (2) the range of spectral type over which the TiO index is useful; and (3) how "pure" the TiO band strength is as an indicator of the temperature, viz., how free it is from sensitivity to other parameters.

5.6.1 Comparison of 8c and MK Temperature Classes

The temperature classes derived from 8c observations of Keenan classification standards are given in Table 2 for comparison with their MK types. The resolution of the MK system for M stars is one-quarter of a subtype; that is, Keenan (1963) considered a quarter-subtype to be the smallest difference he could discern on good-quality classification spectrograms, and, accordingly, his subtypes are divided into four subdivisions, e.g., M2, M2+, M2.5, M3−, M3. For purposes of numerical comparisons with the 8c decimal subdivisions, I have interpreted Keenan's subdivisions as, e.g., M2.0, M2.2, M2.5, M2.8, M3.0. Then taking the differences without regard to sign between the MK and 8c types for the 50 calibration stars with types in the interval K4.0–M6.0, we find a mean difference of 0.24 subtype – exactly what Keenan considered to be the error of the MK types themselves. Indeed, some of the larger differences are likely due to variability, because most of the M4–M6 calibration stars are known to be small-range variables and the MK types are based on single observations at random phases. It is clear that the 8c system reproduces the MK temperature classes for M-type giants very well, and that the small departures from perfect replication could be due almost entirely to uncertainties in the MK classifications.

Does the same conclusion hold for single 8c observations, and for random program stars? For multiply-observed calibration stars, I have examined the 8c classifications obtained from the individual observations and find a mean error of ±0.1 subtype, becoming larger only for known variable stars. The same accuracy should hold for random program stars provided that at least 10,000 counts were recorded at each filter to minimize the effects of photon noise.

A word should be said about the semantically awkward interval from K5 to M0. I have followed the convention employed by Keenan (1963), according to which the types K6, K7, K8, and K9 simply do not exist. The interval from K5 to M0 is the same size as the interval from K4 to K5, or from M0 to M1, and each of these intervals is decimally subdivided in the 8c system. Stars considered to be midway between K5.0 and M0.0 are called K5.5 by Keenan (1963), and so are they still on the 8c system. However, those same stars are called K7 III by Morgan and Keenan (1973), and three such types appear in Table 2. Keenan explained to me (private communication, 1973) that Morgan preferred K7 to K5.5 for these stars in order to emphasize that spectral types are *labels*, not *numerals*. I have ignored this revision so that the interval from K5.0 to M0.0 can be decimally subdivided.

5.6.2 Range of Applicability

Temperature classes cannot be given from the 8c photometry unless measurable absorption by TiO is present. Useful color temperatures can of course be given for stars without TiO (e.g., early-type stars, or carbon stars) from blackbody fits to the continuum points, but these are "photometric" quantities, not spectral classifications. The warmest stars showing measurable TiO absorption are called K3.5 on the 8c system, where the TiO index is about 0.03 mag. Decimal subdivisions start at K4.0, where the index is 0.05 mag.

The TiO absorption at 7,120 Å increases steadily until at least type M6.0, where the index has the value 1.80 mag. At still later types, it is not clear if the absorption continues to grow; this cannot be ascertained from the 8c photometry because the continuum is no longer well defined. Absorption by the VO molecule, which is measured by filter 6 (10,560 Å) after about type M6.0, is also present in filter 2, thereby spoiling that continuum point. As VO is extremely sensitive to temperature and is well measured at filter 6 (where it is referred to a good continuum point at filter 5), I prefer to base temperature classes in the interval M8.0–M10 upon VO alone. In the intermediate interval M6.0–M8.0 – in which VO is present but weak, and TiO is strong but not well measured – I assign types by looking at the indices of both molecules, and combining them in an admittedly subjective manner.

Although the 8c spectral-type scale has been defined entirely on the basis of MK-classified giant stars of luminosity class III, there is no problem extending the scale to the other luminosity classes. M-type dwarfs and supergiants also have spectra dominated by TiO absorption, and because the 8c TiO index is corrected for minor contamination by CN, M stars of all luminosities can be placed on the same scale of spectral types by means of this single index. Classifications on the 8c system have been published for M-type supergiants by White and Wing (1978), and data for M dwarfs are being prepared for publication by Wing et al. (2011).

The stars of type S present special problems. They show enhanced CN strengths (similar to those of M supergiants), and their TiO strengths can be anywhere from zero to very strong. Furthermore, some of the 8c filters can be affected, in the cooler S stars, by bands of LaO and/or ZrS (Hinkle et al. 1989). The 8c system is simply not well suited to handle the great variety of spectra shown by the S stars. Piccirillo (1976, 1977) was able to study the S stars by replacing two of the filters of the 8c system with new filters designed to measure bands of ZrO and LaO explicitly. Ultimately, however, he faced the same problem as spectroscopic observers (Keenan 1954; Keenan and Boeshaar 1980): one can list the strengths of several features, but it is not at all obvious how to distill this information into a two-dimensional (temperature/luminosity) classification.

5.6.3 Sensitivity of TiO to Temperature

We have already noted that the TiO band strength correlates well with atomic-line ratios that are known to be temperature-sensitive. But a more direct – and quantitative – way to measure the sensitivity of TiO to temperature is to examine the relation between TiO bandstrength and continuum color. To do this, we need a sample of stars whose colors are not affected by interstellar reddening. The calibration stars of Table 2 will serve this purpose, as they all lie within about 100 pc of the Sun.

In Fig. 3, the TiO indices of the MK calibration stars of Table 2 are plotted against a continuum color index, formed from the magnitudes at the best continuum points (filters 2 and 6). The relation is essentially linear all the way from type K5.0 (where the feature depth is 15%) to type M5.0 (where the continuum is 4 times as bright as the feature). After type M5, the relation begins to flatten out, probably because of molecular absorption creeping into the bandpass of filter 2; the four calibration stars of latest type are off-scale to the right for this reason.

The width of the relation is believed to be intrinsic, because these stars are very well observed (the photometric errors are no larger than the dots) and because interstellar reddening (which moves stars to the right in this diagram) should be insignificant for this sample. Furthermore, stellar variability plays no role here, as the two quantities plotted were obtained from the same 8c data sets.

In order to quantify the width of this relation, I have labeled three of the stars: the M2.0 III star χ Peg sits on the upper envelope of the relation; λ Aqr has the same TiO strength but a redder color; and κ Ser has the same color as χ Peg but weaker TiO bands. The 8c spectra of these stars, shown in Fig. 4, do not seem in any way peculiar: they have similar CN strengths and show no sign of absorption other than the usual TiO and CN. Thus we interpret the width of the TiO/color relation as indicating the range in TiO strength that can occur at a given temperature, for otherwise normal solar-neighborhood giants.

For χ Peg and κ Ser – stars of the same temperature but very different TiO strength – we find TiO indices of 0.49 and 0.27 mag, respectively. Thus χ Peg has nearly twice as large a TiO index as κ Ser, and because both stars fall on the linear part of the TiO–color relation where the TiO band is evidently not saturated, we conclude that χ Peg has nearly twice the number of TiO molecules above its photosphere. Whether this is due to a difference in metallicity affecting the Ti abundance,

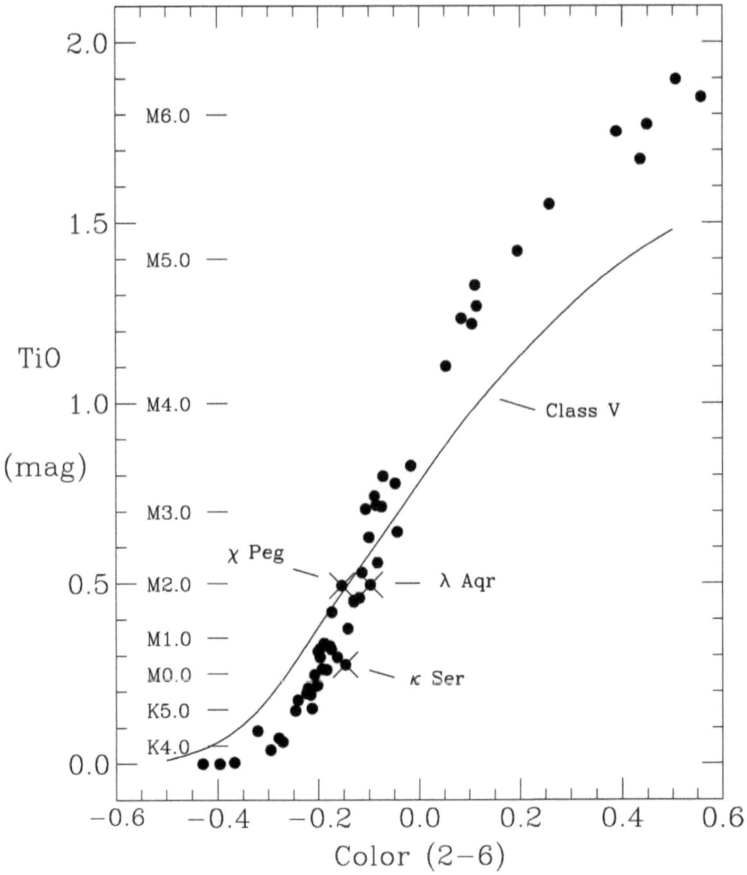

Fig. 3 The relation between TiO index and continuum color for a sample of unreddened stars, namely the K and M giant calibration stars of Table 2 (*dots*). For comparison, the mean relation for luminosity class V dwarfs is also shown. The calibration of the TiO index in terms of spectral type is shown at the left. The (2–6) color is formed from the magnitudes at filters 2 and 6, which are good continuum points in both giants and dwarfs. The three stars marked with *crosses* are discussed in the text

or to a difference in the O/C ratio, or to a physical difference in, say, the gas pressure or microturbulence, we cannot say without further input.

Now by comparing χ Peg and λ Aqr – which have the same band strength despite having different temperatures – we can evaluate the width of the TiO–color relation in terms of temperature. These two stars differ by 0.057 mag in C(2–6), and by 145°K in color temperature. Thus we can say that the width of the relation is approximately 150°K. If we are using the measured TiO index to judge the intrinsic color temperature of a mid-M star, the probable error of the determination, taken to be half the width of the intrinsic relation, is about 75°K.

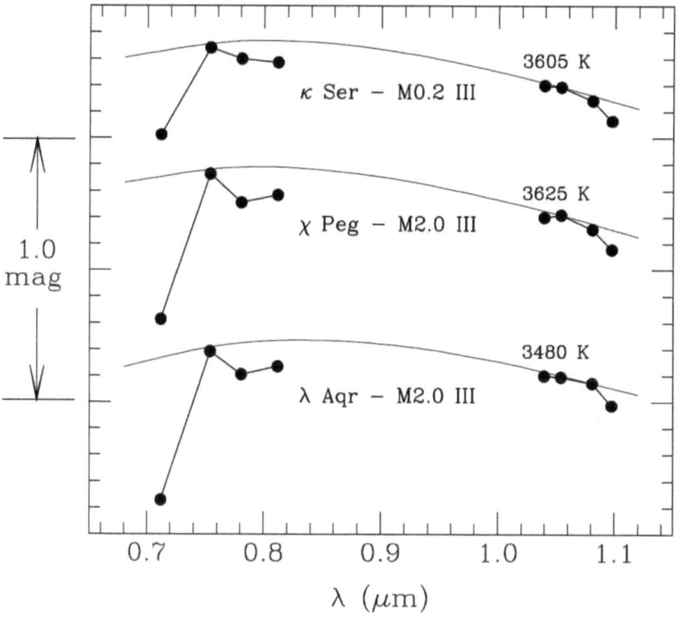

Fig. 4 Eight-color spectra of the three stars labeled in Fig. 3, illustrating the intrinsic width of the TiO–color relation

In any discussion of "accuracy," we must be careful to define what we are talking about. If the purpose of the TiO measurement is simply to determine the spectral type (i.e., the TiO strength), then the accuracy of the 8c measurement is very high, because the TiO index places the spectral type on Keenan's MK scale with an error of no more than ± 0.1 subtype. But if the purpose in measuring TiO is to judge the temperature, then the width of the intrinsic TiO–color relation comes into play and the error is approximately 10 times larger – about 75°K, which corresponds to nearly a full subtype.

The color temperatures discussed here, obtained from blackbody fits to the data, are of course not the same as effective temperatures. The difference can be found by computing synthetic 8c photometry from model atmospheres. A preliminary study by Wing et al. (1985) found that, for stars in the range K0 III – K4 III, effective temperatures are higher than the 8c color temperatures by about 200°K. The difference is due almost entirely to the effect of H^- opacity and varies only slightly with temperature in this range. The difference is therefore an offset which in principle can be computed almost exactly, without adding significantly to the uncertainty of temperature determinations.

5.6.4 ***In*sensitivity of TiO to Luminosity**

The concentration of any molecule in a stellar atmosphere depends upon the temperature, gas pressure, and composition, while the strength of any spectral feature depends also upon the opacity at that wavelength. We have seen that the strengths of the optical TiO bands are very sensitive to temperature; what can we say about their sensitivity to other parameters? A perfect classification criterion would be sensitive to one parameter only, but realistically we cannot expect a molecular band strength to have that property.

M stars of different luminosity classes differ enormously in size, and hence also in the gas pressures of their photospheres. Therefore, a comparison of TiO strengths of M-type dwarfs and giants of the same temperature should provide a good test of the sensitivity of TiO to gas pressure, or luminosity.

Narrow-band photometry on the eight-color system is available for more than 200 dwarf stars with spectral types between K4 V and M6 V (Wing et al. 2011). These stars are all among our "nearest neighbors" (mostly within 10 pc of the Sun), so that we need not be concerned about interstellar reddening. A plot of TiO index vs. (2–6) continuum color, similar to Fig. 3 for the giant calibration stars, has been constructed for these luminosity class V stars, and the mean relation has been imported into Fig. 3.

It is clear that the TiO–color relations for giants and dwarfs have different slopes. But what I think is most remarkable is how similar the two relations are: instead of being greatly displaced from one another as we might expect from the very different gas pressures involved, the two relations intersect near the middle of the diagram. Indeed, at around spectral type M2 or M3, giants and dwarfs of the same *color* temperature have the same band strength of TiO. While giants and dwarfs of the same C(2-6) color temperature may not have the same *effective* temperature, the difference is not expected to be large. As noted above, the offset is approximately 200°K in giants, and it should be similar in dwarfs because in both cases it is the result of the wavelength dependence of the bound-free opacity of the H^- ion.

A model-atmosphere analysis which will allow more definite conclusions is being carried out in collaboration with U. G. Jørgensen of the Niels Bohr Institute. A preliminary interpretation can be given here, however. In the range K4–M1, dwarfs have stronger TiO bands than giants of the same temperature as the expected direct result of their higher gas pressure; but with decreasing temperature, the band strengths in giants grow at a greater rate than in dwarfs because the H^- opacity at 7,540 Å (filter 2, where the continuum is measured) decreases faster in giants than in dwarfs.

Despite these differences, the main conclusion must be that the measured TiO band strength is almost entirely a reflection of the temperature, varying very little with luminosity over a huge range in gas pressure. In other words, for purposes of spectral classification, TiO provides a nearly perfect criterion for temperature.

I have not discussed supergiants, primarily because all M supergiants are reddened by interstellar dust to some degree and their intrinsic colors are rather uncertain. It does appear, however, that M supergiants of a given spectral type (TiO strength) have somewhat redder intrinsic colors than M giants of the same spectral type. This result has been obtained for the Iab supergiants in the Double Cluster h & χ Persei, which were individually de-reddened with the aid of *UBV* photometry of early-type stars in the vicinity of each (Wildey 1964); see the discussion of Fig. 3

of MacConnell et al. (1992). Still, we do not err greatly if we think of M-type giants and supergiants as having similar temperatures at any given TiO band strength.

5.7 CN as an Indicator of Luminosity

Prior to about 1965, virtually all discussions of CN bands in stellar spectra referred to the Violet ($B^2\Sigma$ – $X^2\Sigma$) System, the (0,0) bandhead of which lies at 3,883 Å, just shortward of the Ca II K-line. The Red ($A^2\Pi$ – $X^2\Sigma$) System, however, has bands throughout the visual and near-infrared regions and is more often the subject of modern investigations. Both systems arise from the molecule's ground electronic state and so would be expected to show similar behavior. However, because of a number of extraneous effects (blends, opacity effects, and changes of continuum slope caused by interstellar dust or detector limitations), the observed behavior of the two band systems is rather different.

5.7.1 Historical Background: Observations of Violet System Bands

The positive luminosity effect of CN bands in stars of types G and K was first reported by Lindblad (1922), who observed the (0,0) and (0,1) bands of the Violet System which degrade to shorter wavelengths from heads at 3,883 Å and 4,216 Å, respectively. When the (0,0) band is strong, it can cause the observed spectrum to be cut off at the short-wavelength end, and Lindblad (1919) was already using the minimum wavelength of spectra of G and K stars – as seen on very short objective-grating spectra – as an absolute-magnitude indicator before the cause of the cut-off had even been identified as CN. Lindblad was concerned simply with distinguishing giant stars from dwarfs, a distinction that was then normally made on the basis of proper motions but for which spectroscopic criteria were being developed at Mount Wilson (Adams et al. 1926). By 1922, Lindblad was able to report that the CN bands were strong in all G and K giants that he had observed, whereas "in dwarf stars of all types the bands are only faintly developed."

Because the region depressed by the $\Delta v = -1$ sequence of CN is rather broad – the deepest part is usually described as 4,144–4,184 Å – it can be seen easily at dispersions too low to allow the use of the usual atomic-line criteria. During the 1930s it was widely used for distinguishing giants from dwarfs in low-dispersion spectroscopic surveys (e.g., by Lindblad and Stenquist 1934; Iwanowska 1936; Becker 1938). Results for supergiants – sometimes called "super-giants" or "pseudo-Cepheids," the high luminosities of which had only recently been established (Payne 1930) – were ambiguous, primarily because the steep slopes of their (reddened) continua often made the CN feature hard to recognize; this problem was specifically discussed by Hoffleit (1939). A careful study of the applicability of the CN luminosity criterion in the case of supergiants was carried out by Keenan (1941), who showed that supergiants in the range G4–K1 have clearly enhanced CN bands as compared to giants. In the later K subdivisions, however, Keenan found that the CN (0,1) band becomes lost among atomic lines, at all luminosities.

In the original MK Atlas (Morgan et al. 1943), I find no mention of the CN molecule. However, on Plate 41 ("Luminosity Effects At G5") the authors note a "change with luminosity in the appearance of the three broad blends between λ4144 and λ4215," and on Plate 48 ("Luminosity Effects At K3") they comment that "Absolute magnitude differences are also shown by the ratio of the intensity of the continuous spectrum on each side of λ4215." These effects are, in fact, caused by the $\Delta v = -1$ sequence of CN.

Several developments during the 1940s and 1950s caused the CN molecule to be regarded in a different light. With the recognition of distinct stellar populations in the Galaxy (Baade 1944), and Roman's (1950) discovery that high-velocity stars have systematically weak atomic lines, it became clear that significant abundance differences could be found among, say, the common G and K giants. High-velocity stars were found to have systematically weaker CN than low-velocity stars of the same temperature and luminosity, although exceptions were also found. Also, some strong–CN stars were found that did not appear to be over-luminous. Thus the CN molecule became less interesting as a classification criterion, but perhaps more interesting in connection with stellar evolution, abundances, and populations.

Using a grating spectrometer and photomultiplier tubes, Griffin and Redman (1960) surveyed the behavior of CN in more than 700 late-type stars. They measured the $\Delta v = -1$ sequence with a 50 Å bandpass (4,164–4,214 Å), with simultaneous measurements in two sidebands. Their pioneering study – the first of a series from Cambridge University – demonstrated the great potential of narrow-band photometry in the measurement of strong spectral features, even under poor climatic conditions. They confirmed the positive luminosity effect of the blue CN feature, but noted that the absolute magnitudes they computed from CN strength for giant stars (luminosity classes II and III) did not correlate very closely with luminosities computed from the MK luminosity subdivisions, the mean difference being about 1.5 mag. They were able to detect CN absorption to somewhat later types than had been possible spectroscopically; even so, they found that the blue CN feature becomes lost in the "noise" of atomic lines at about type M0.

In his comprehensive review of spectral classification, Keenan (1963) gives considerable attention to the blue CN feature, and in particular to the results of Griffin and Redman. Keenan's conclusion is that, "As luminosity indicators, the cyanogen bands remain valuable because of their great sensitivity, but they cannot safely be used alone – other criteria should always be observed as checks."

5.7.2 Measurements of Red System Bands

Several bands of the Red System of CN – the (2,0) band near 8,100 Å, the (1,0) band near 9,200 Å, and the (0,0) band near 11,000 Å – were measured by the writer as part of a survey of the near-infrared spectra of late-type stars (Wing, 1967). A spectrum scanner was used with a 30 Å bandpass to define a 27–color photometric system on which several hundred stars were observed over the course of two years. In two respects, the bands of the Red System were found to behave like their Violet System counterparts: (1) they unambiguously separate giants from dwarfs, and supergiants from giants; and (2) their strengths among luminosity class II and III giants do not show a detailed correlation with other indicators of luminosity. That is to say, strong-CN giants are not necessarily more luminous than weak-CN giants. But with respect to dependence on temperature, the Red System bands show a different behavior: they remain visible and measurable to much later spectral types than do the Violet System bands. At any given luminosity class, the Violet bands reach a maximum strength in the early K subtypes and then decline to invisibility at about M0, while the Red bands, after reaching a similar maximum, maintain their strength at a nearly constant level well into the M subtypes, to at least M4 or M5. This is important for classification purposes: not only are the Red System bands measurable over a wider range of spectral type, but their observed strengths are nearly independent of temperature for types later than early K.

How can we understand the lack of dependence on temperature shown by the Red System CN bands, while the Violet System bands depend strongly upon both luminosity and temperature? Calculations of molecular dissociation equilibria show that, for temperatures below about $4,000°$K, the number of CN molecules above the photosphere should decrease with decreasing temperature as atoms of both C and N become increasingly bound into other molecules. However, the opacity due to the negative ion H^- also decreases as atoms of metals become neutral and the supply of free electrons decreases; evidently in the red and near-infrared, where bound-free opacity by H^- is strongest, these effects roughly cancel one another, so that the band strengths of Red System CN bands stay approximately constant, at a given gas pressure, over a wide range of temperature. This cancellation does not occur for the bands of the Violet System because H^- is less important at the shorter wavelengths, and its place is increasingly taken by the overlapping wings of innumerable atomic lines. In fact, the total opacity near 4,200 Å probably increases with decreasing temperature, while the opacity in the 8,000–11,000 Å region decreases.

The CN index on the eight-color system, defined as the mean of the numerically–similar depressions at filters 4 and 8 (Equation 10), is given in Table 2 for the 60 MK calibration stars. These stars were selected as having similar MK luminosities, and we see that they also have similar CN indices, nearly all of them within the range 60–120 mmag, with no discernable trend against temperature class. Because of this flatness, it is possible to assign luminosity classes strictly on the basis of the numerical value of the CN index, over the entire range of spectral type for which the eight-color system can assign TiO-based temperature classes, approximately K4–M6. Luminosity classes are currently assigned according to the values listed in Table 3.

Several comments should be made concerning the luminosity classes assigned on the eight-color system:

- Luminosity class IV does not exist for stars of late enough type to show TiO bands (as on the MK system)

Table 3 CN–Based Luminosity Classes

CN index (mag)	Assigned luminosity class
0.000 – 0.034	V
0.035 – 0.100	III
0.101 – 0.124	II
0.125 – 0.174	Ib
0.175 – 0.244	Iab
>0.245	Ia

- The borderline between classes V and III is set at 0.035 mag because no well-observed dwarf has been found with $CN > 0.03$
- Stars classified as class III giants on the MK system show a wide range of CN strengths; the MK subdivisions IIIa, IIIab, IIIb do *not* correlate with CN strength, and all stars with $0.04 < CN \leq 0.10$ are simply called III on the 8c system
- Most stars of MK class II have CN in the range 0.100–0.124; however, some stars of MK class III also fall in this range. The 8c system cannot distinguish between strong–CN giants and the somewhat more luminous "bright giants" of class II
- All MK-classified supergiants of types G, K, and M show strong CN bands ($CN \geq 0.125$). The MK subdivisions Ia, Iab, Ib *do* correlate with CN strength; the CN–based luminosity subdivisions for supergiants reproduce the MK subdivisions assigned by Keenan almost perfectly

This last point is important. The MK subdivisions of luminosity class I, based entirely on atomic-line ratios, agree with the CN–based subdivisions given by the 8c photometry and the borderlines specified in Table 3. The same *cannot* be said for luminosity class III giants, for which the MK subdivisions are not correlated – or perhaps are even slightly anti-correlated – with CN strength. The difference is undoubtedly due to the fact that all supergiants are young, Population I objects, whereas the giants embrace a wide range of ages and population types. The implication seems to be that late-type supergiants of classes Ib and brighter – in our Galaxy, at least – have very similar chemical compositions, whereas a range in metallicity and/or CNO abundances is encountered among the class III giants of the solar neighborhood.

A set of K and M supergiants belonging to the Small Magellanic Cloud have been observed on the 8c system (Wing et al. 2004), and the results indicate that a different calibration of the CN index is needed for this metal-poor sample.

Luminosity classes given from TiO/CN classification photometry are limited to the divisions listed in Table 3. Specifically, no attempt is made to distinguish subdwarfs from dwarfs (neither of which have significant CN absorption) or to subdivide the class III giants (which are subject to composition differences).

5.8 Distances from TiO/CN Photometry

An attractive feature of the 8c system is that it provides both photometric quantities (a near-infrared magnitude and a continuum color) and spectroscopic quantities (reddening-free indices of molecular band strengths, leading to a two-dimensional spectral classification). It is therefore possible to compute distances from the 8c photometry alone. Reddening is determined by plotting a star on a TiO–color diagram such as Fig. 3; this requires knowledge of the intrinsic relation for the appropriate lumunosity class. A calibration of the two-dimensional spectral classifications in terms of absolute magnitude is also needed. For this purpose we use the absolute magnitude $M(104)$ at 10,400 Å (filter 5), for comparison with the observed $I(104)$ magnitude. The method, which is most effective when applied to M supergiants, is described by MacConnell et al. (1992).

6 Final Remarks, and a Suggestion

In the course of writing (and re-writing) this chapter, I came to realize that one cannot discuss the accuracy attainable in the classification of stellar spectra through use of photoelectric photometry without giving considerable attention to the possibility that "photometric types," even when very precise, may sometimes simply be wrong – i.e., at odds with classifications given by inspection of the spectrum. I have therefore given examples of the distinction between "photometric classifications" and "true spectral classifications," and I have discussed the criteria I would use to make that distinction. Briefly, if a classification is based on photometry of any kind, it should in most cases be called a "photometric" classification, and it should be called a "spectral" classification only if it involves the measurement of specific spectral features. Going one step further, it could be argued that the classification should not be presented in MK notation unless those spectral features are separately sensitive to temperature and luminosity.

But does this distinction really matter? In many applications, a classification would be used in exactly the same way, regardless of its origin. Some systems of photometric classification have a high probability of being "correct," in the sense that they provide the same result as a conventional MK classification. Also, in statistical applications it is probably possible to allow statistically for the small fraction of photometric types that are incorrect. Even so, I feel that the user of a classification should always be able to tell whether or not the spectrum has actually been observed.

Returning to our first, rather trivial, example (Sect. 2.2) of a spectral classification reported on the basis of an observed $B-V$ color: the 1 % accuracy of the color index should yield a spectral type considerably more precise than can be obtained from visual inspection of a spectrogram, but there is a substantial probability that this photometric classification is simply wrong (e.g., if the star is reddened, or chemically peculiar, or of a different luminosity than expected). I think we would all agree that a photometric classification based on the observation of a color such as $B-V$ or $V-K$ is not a spectral classification at all, and should never be published as such. But now consider the carefully-designed systems of multicolor intermediate–bandwidth filter photometry that were discussed in Sect. 4. They have a much lower probability of being wrong because they offer several checks: the reddening can be determined (at least for certain kinds of stars), the luminosity class can be checked, and certain kinds of chemical peculiarity can be recognized. Even so, if the position of a star in a diagram intended to determine the reddening is displaced by the unexpected strength of certain spectral features, the derived reddening will be wrong and so will be the resulting classification.

What is the photometrist supposed to do? I have heard MK classifiers express the opinion that classifications should not be given in the MK notation unless they have been obtained by the "MK process." This purist position becomes increasingly difficult to maintain, however, as even spectroscopic classifications are increasingly done with different detectors, wavelength regions, and resolutions. The photometrist is not likely to want to invent an entirely new system of nomenclature; the MK notation has the great advantage that it is widely understood and has been used for more than half a century.

My suggestion is addressed to compilers of catalogues, or other compilations in which photometric types and spectral types may be mixed together. I propose that photometric types in the MK notation be identified with the simple prefix "p". Thus a star photometrically classified as a K2 dwarf or an F0 Ia supergiant would appear in catalogues as pK2 V and pF0 Ia. The catalogue user would then know whether the spectrum had been observed, and could include such types or not, according to the application at hand. No other change would be needed; photometrists could continue to use the MK notation, and the same calibrations would apply.

I do not, however, propose using the "p-prefix" for classifications from the 8c narrow-band system of TiO/CN photometry, because these meet the criteria for true spectral classifications.

Acknowledgements I have benefitted from numerous opportunities to use the excellent NOAO facilities on Kitt Peak and Cerro Tololo over the years, and I thank the corresponding Directors, support staffs, and TACs for making that possible. I also thank my colleagues in IAU Commission 45 – both spectroscopists and photometrists – for stimulating discussions on many of the topics considered here.

References

Adams, W. S., Joy, A. H., & Humason, M. L. (1926). *Astrophysical Journal*, **64**, 225
Baade, W. (1944). *Astrophysical Journal*, **100**, 137
Becker, W. (1938). Potsdam Observatory Pub., **28**(3)
Bessell, M. S. & Brett, J. M. (1988). *Publications of the Astronomical Society of the Pacific*, **100**, 1134
Cannon, A. J., & Pickering, E. C. (1914). *Introduction to the Henry Draper Catalogue*, Vol. **1**; *Annals of Harvard College Observatory*, Vol. **91**
Crawford, D. L. (1958). *Astrophysical Journal*, **128**, 185
Crawford, D. L. (1975). *Astronomical Journal*, **80**, 955
Crawford, D. L. (1978). *Astronomical Journal*, **83**, 48
Crawford, D. L. (1979). *Astronomical Journal*, **84**, 1858
Cutri, R. M., et al. (2003). *2MASS All Sky Catalog of Point Sources*, (Pasadena: IPAC/Caltech)

Frogel, J. A., Persson, S. E., & Cohen, J. G. (1983). *Astrophysical Journal, Supplement*, **53**, 713
Golay, M. (1963). Publ. Observatory Genève, **64**, 419
Golay, M. (1974). *Introduction to Astronomical Photometry*. (Dordrecht: D. Reidel Publishing)
Griffin, R. F., & Redman, R. O. (1960). *Monthly Notices of the Royal Astronomical Society*, **120**, 287
Hinkle, K. H., Lambert, D. L., & Wing, R. F. (1989). *Monthly Notices of the Royal Astronomical Society*, **238**, 1365
Hoffleit, D. (1939). *Astrophysical Journal*, **90**, 621
Iwanowska, W. (1936). *Stockholms Observatoriums Annaler*, **12**, pp. 5.1–5.26
Johnson, H. L. (1958). *Lowell Observatory Bulletin*, **4**, 37
Johnson, H. L. (1966). *Annual Review of Astron and Astrophys*, **4**, 193
Johnson, H. L. (1968). In B. M. Middlehurst & L. H. Aller (eds.), *Nebulae and Interstellar Matter*, (Chicago: University of Chicago Press), p. 167
Johnson, H. L., & Morgan, W. W. (1953). *Astrophysical Journal*, **117**, 313
Jørgensen, U. G., & Wing, R. F. (2000). In *IAU symposium 177: The Carbon Star Phenomenon*, (Dordrecht: Kluwer), p. 543
Keenan, P. C. (1941). *Astrophysical Journal*, **93**, 475
Keenan, P. C. (1954). *Astrophysical Journal*, **120**, 484
Keenan, P. C. (1963). In K. Aa. Strand (ed.), *Basic Astronomical Data*, (Chicago: University of Chicago Press), p. 78
Keenan, P. C., & Boeshaar, P. C. (1980). *Astrophysical Journal, Supplement*, **43**, 379
Keenan, P. C., & Pitts, R. E. (1980). *Astrophysical Journal, Supplement*, **42**, 541
Landi Dessy, J., & Keenan, P. C. (1966). *Astrophysical Journal*, **146**, 587
Lindblad, B. (1919). *Astrophysical Journal*, **49**, 289
Lindblad, B. (1922). *Astrophysical Journal*, **55**, 85
Lindblad, B., & Stenquist, E. (1934). *Stockholms Observatoriums Annaler*, **11**(12)
MacConnell, D. J., Wing, R. F., & Costa, E. (1992). *Astronomical Journal*, **104**, 821
McClure, R. D. (1970). *Astronomical Journal*, **75**, 41
McClure, R. D. (1979). In A. G. D. Philip (ed.), *Problems of Calibration of Multicolor Photometric Systems*, Dudley Observatory Report No. 14, p. 83
McClure, R. D., & van den Bergh, S. (1968). *Astronomical Journal*, **73**, 313
Milone, E. F., & Young, A. T. (2005). *Publications of the Astronomical Society of the Pacific*, **117**, 485
Morgan, W. W., Code, A. D., & Whitford, A. E. (1955). *Astrophysical Journal, Supplement*, **2**, 41
Morgan, W. W., & Harris, D. L. (1956). *Vistas in Astronomy*, (New York: Pergamon), Vol. **2**, p. 1124
Morgan, W. W., & Keenan, P. C. (1973). *Annual Review of Astron and Astrophys*, **11**, 29
Morgan, W. W., Keenan, P. C., & Kellman, E. (1943). *An Atlas of Stellar Spectra*, (Chicago: University of Chicago Press)
Morgan, W. W., Whitford, A. E., & Code, A. D. (1953). *Astrophysical Journal*, **118**, 318
Neugebauer, G., & Leighton, R. B. (1969). *Two-Micron Sky Survey*, NASA SP-3047
Payne, C. H. (1930). *The Stars of High Luminosity. Harvard Observatory Monograph* No. 3, (New York: McGraw-Hill Book Company)
Piccirillo, J. (1976). *Publications of the Astronomical Society of the Pacific*, **88**, 680
Piccirillo, J. (1977). *Bulletin of the American Astronomical Society*, **9**, 294
Roman, N. G. (1950). *Astrophysical Journal*, **11**, 554
Spinrad, H., & Wing, R. F. (1969). *Annual Review of Astron and Astrophys*, **7**, 249
Straižys, V. (1963). Bulletin of the Vilnius Astronomical Observatory, **6**, 1
Straižys, V. (1979). In A. G. D. Philip (ed.), *Problems of Calibration of Multicolor Photometric Systems*. Dudley Observatory Report No. 14, p. 215
Straižys, V. (1992). *Multicolor Stellar Photometry*, (Tucson: Pachart Publishing House)
Strömgren, B. (1963). In K. Aa. Strand (ed.), *Basic Astronomical Data*, (Chicago: University of Chicago Press), p. 123
Strömgren, B. (1966). *Annual Review of Astron and Astrophys*, **4**, 433

White, N. M., & Wing, R. F. (1978). *Astrophysical Journal*, **222**, 209
Wildey, R. L. (1964). *Astrophysical Journal, Supplement*, **8**, 439
Wildey, R. L., Burbudge, E. M., Sandage, A. R., & Burbidge, G. R. (1962). *Astrophysical Journal*, **135**, 94
Wing, R. F. (1967). PhD dissertation, University of California, Berkeley
Wing, R. F. (1971). In G. W. Lockwood & H. M. Dyck (eds.), *Colloquium on Late-Type Stars*, (Tucson: Kitt Peak National Observatory), p. 145
Wing, R. F. (1979). In A. G. D. Philip (ed.), *Problems of Calibration of Multicolor Photometric Systems*, Dudley Observatory, Report No. 14, p. 499
Wing, R. F. (1994). *Revista Mexicana de Astronomia y Astrofisica*, **29**, 175
Wing, R. F. (2005). *Bulletin of the American Astronomical Society*, **37**, 1442
Wing, R. F. (2009). *Bulletin of the American Astronomical Society*, **41**, 200
Wing, R. F., Gustafsson, B., & Eriksson, K. (1985). In D. S. Hayes, L. E. Pasinetti & A. G. D. Philip (eds.), *IAU Symposium 111: Calibration of Fundamental Stellar Quantities*, (Dordrecht: D. Reidel Publishing), p. 571
Wing, R. F., & Jørgensen, U. G. (2003). *Journal of the American Association of Variable Star Observers*, **31**, 110
Wing, R. F., Dean, C. A., & van der Bliek, N. S. (2011), *in preparation*
Wing, R. F., Walker, K. M., MacConnell, D. M., & Costa, E. (2004). In D. W. Kurtz & K. R. Pollard (eds.), *IAU Colloquim 193, Variable Stars in the Local Group*, (San Francisco: ASP) ASP Conference Series, Vol. **310**, p. 317

Absolute Photometry: Past and Present

Martin Cohen

1 Introduction

I will try to cover the history of absolutely calibrated phometry in astronomy, spanning the wavelength range from Far-Ultraviolet (FUV) to Mid-Infrared (MIR). Such an agenda will of necessity cite the direct comparisons of stars with fiducial sources and hardware devices. These include standards that are traceable to primary laboratory references, lamps, black bodies, and attenuators as used on the ground, or in space on brief rocket flights, or on long-lived satellites. Either these devices are themselves deployed or a deployable version is compared directly with a primary standard prior to, and/or following observations.

The purpose of such comparisons is to lead to the establishment of standard stars, which imply ideally that these are predictable, easy to model, nonvariable sources. The role of stellar models, or rather, of their associated synthetic spectra, constitutes a vital but somewhat changeable intermediary that is intended to help us to interpret the similarities and differences between what we observe and what we had expected to observe. I'll break down the material by wavelength coverage but will touch solely on the highlights of this long and fascinating history.

2 Extreme UV (EUV) to Near-UV (NUV): 100–3,350 Å

For more than a decade space missions were flown to establish stellar standards or to intercompare stellar energy distributions with predicted stellar synthetic spectra of hot stars. The on-board fiducial devices were photomultipliers which had been used to observe an NBS characterized photodiode or a synchrotron source. The Apollo-17 S-169 Experiment (Henry et al. 1975) secured spectra of six OB dwarfs with an error of $\pm 10\%$. Rocket flights were carried out by Bless et al. (1976) who observed

M. Cohen (✉)
University of California, Berkeley, CA 94709, USA
e-mail: mcohen@astro.berkeley.edu

E.F. Milone and C. Sterken (eds.), *Astronomical Photometry: Past, Present, and Future*, Astrophysics and Space Science Library 373, DOI 10.1007/978-1-4419-8050-2_8,

α Leo, α Vir and η UMa, with a quoted accuracy of ±10%. The first extrasolar EUV source to be observed was Hz43, by Lampton et al. (1976), who achieved an accuracy of ±6%. Strongylis and Bohlin (1979) compared results obtained from rockets with those from the OAO2 satellite. In spite of the stated formal uncertainties associated with these measurements, these authors found discrepancies of order 35% in the region between 1,200 and 1,600 Å. Bohlin et al. (1980) also noted 10% differences longward of 1,200 Å between different measurements of the IUE low-dispersion calibrator η UMa. This star was also used for the calibration of Voyager 1 and 2 spectra beyond 1,250 Å (Holberg et al. 1982) and for that of the TD1 satellite (Carnochan 1982). Holberg and colleagues noted that, while Voyager and data by Brune et al. (1979) and Carruthers et al. (1981) were essentially in accord near 1,200 Å, the continuum observed by the Voyagers from 912 to 1,200 Å was about 60% higher than these other measurements. Holberg et al. (1982), therefore, advocated a new approach. They proposed an adjusted Voyager calibration based on the adoption of Kurucz (1979) LTE models for eight hot stars normalized at 5,500 Å to α Lyr and to the DA White Dwarfs (WDs) Hz43 and G191-B2B. Additional changes in calibration followed when Bohlin (1990,1995,1996) replaced the $0^{\mathrm{m}}.06$ variable B3V "standard" star, η UMa, by NLTE models of WDs for the IUE, FOS, STIS, and NICMOS instruments. Holberg and Polidan (1985) also reported that B-dwarfs had been found to be variable at the 5–40% level in the 912–1,200 Å regime, further vindicating these changes of paradigm.

3 DA White Dwarfs

DA WDs had already become the primary FUV calibrators at the ±10% level of precision (e.g., Sing et al. 2002). Simplicity of the spectral modeling for a pure Hydrogen WD was a leading reason for their adoption. These stars are continuum-dominated, radiative objects. To represent their photospheres requires only effective temperature and surface gravity. The minimal influence of extinction on their observed spectra derives from the relative proximity to us of the selected WD calibrators. Holberg and Bergeron (2006) have also demonstrated the consistency, at about the ±1% level, of synthetic photometry performed using these stars. It was suggested that WD models could bridge the range from EUV to MIR. This was supported, for example, by the work of Tremblay and Bergeron (2007 and references therein) who have compared the capability of a large grid of complete spectra of WD models to fit simultaneously the combination of photometry in the optical, near-IR (2MASS), and MIR (Spitzer/IRAC data at 4.5 and 8.0 μm).

4 Vega: Optical/NIR Comparisons with Standard Lamps

In the decades between 1970 and 2000 there were two critical reviews of direct comparisons at 5,556 Å between Vega and standard lamps or blackbodies. That by Hayes (1985) advocated an F(λ) of $3.44 \pm 0.05 \times$ E-12 W cm^{-2} μm^{-1} with an uncertainty of 1.45%. Megessier (1995) preferred a value of $3.46 \pm 0.02 \times$ E-12 W $cm^{-2}\,\mu m^{-1}$ with 0.7% uncertainty. This value was obtained by rejecting data from the Tungsten lamp used in 1969 at Palomar, on the basis of potential systematic errors in its data. Note that the change in adopted flux value between the recommendations of these two critical analyses is equivalent to only 0.6%.

In the NIR, carefully chosen narrow-band filters were used to restrict the influence of water vapor on the direct comparisons of Vega with a furnace, that were undertaken Blackwell et al. (1983) and by Selby et al. (1983). The optics were designed so that both the star and the furnace on the telescope illuminated the detector in the same way. The inclusion of attenuators to avoid detector saturation by the furnace was a further improvement over earlier experimental designs (Selby et al. 1980). The furnace at the telescope was compared with a precision furnace from the UK's National Physical Laboratory with known absolute emissivity. This group achieved 3% precision for Vega's irradiance at 1.24, 2.20, 3.76 µm, and 12% at the terrestially more challenging wavelength of 4.6 µm. Their data indicated an 8% excess over a Vega model by Dreiling and Bell (1980) at 2.20 µm. Subsequent efforts by Mountain et al. (1985) eliminated the attenuators in favor of a spectrometer and measured Vega at 4.92 µm to 7% precision, corresponding to an 11% excess over the same Vega model. These authors concluded (Blackell 1991, private communication) that these were extremely difficult experiments to pursue on the ground and that systematic errors were surely underestimated and posed the ultimate limits to accuracy.

5 Absolute MIR Validation by the Midcourse Experiment (MSX)

The response of the six MSX MIR bands were precisely (<0.5% rms) tied (Price et al. 2004) to the Cohen–Walker–Witteborn (1992b , CWW) fluxes for the Kurucz (1991) Sirius model used as the MSX primary calibrator (parameters: 9,850 K, 4.25, +0.5, 0 km/s) by Cohen et al. (1992a). The SPIRIT III instrument on MSX carried out on-orbit direct, independent, absolute measurements of stars against NIST-traceable black emissive reference spheres at 8.3, 12.1, 14.7, 21.3 µm (Price et al. 2004). These experiments validated the CWW zero-magnitude flux scale (adopted to be the (1991) Kurucz model synthetic spectrum for Vega's photosphere cited by Cohen et al. (1992a with parameters: 9,400 K, 3.90, −0.5, 0 km/s). Averaged over the four long wavelength bands of SPIRIT III from 8.3 to 21 µm, CWW's zero magnitude definitions are accurate to 1.1%, below their 1.45% uncertainties. The absolute fluxes of all of the CWW secondary standards (bright K and M giants) are validated if the flux from the CWW Sirius spectrum is increased by 1.0%, again well within these authors' 1.46% errors in their absolute adopted spectrum of Sirius.

Figure 1 presents direct and independent absolute comparisons of the IR energy distributions of six CWW bright secondary standard stars. Each $\lambda^4 \times F(\lambda)$ plot offers the CWW predicted stellar spectrum. The mean is in red; the ± 1-σ bounds flank the mean in green; and the six MSX absolute photometric measurements ("MSX global calibration") are in blue and indicate both the ± 1-σ flux uncertainties of the

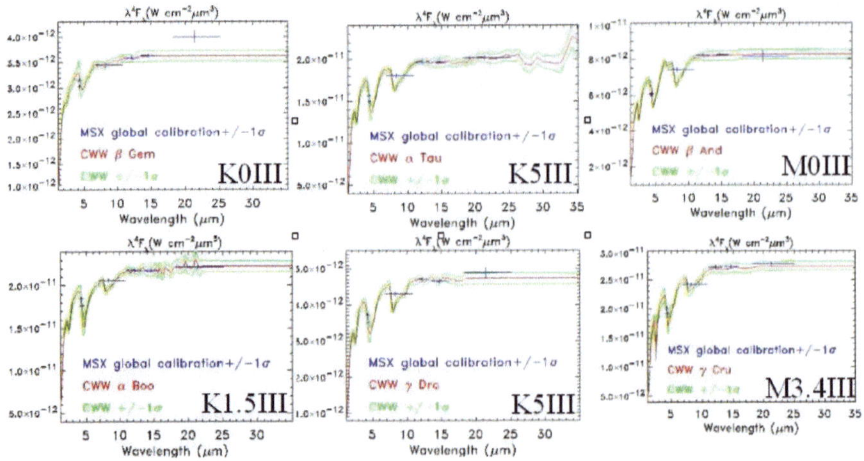

Fig. 1 Direct and independent absolute comparisons of the IR energy distributions of six CWW bright secondary standard stars. Each $\lambda^4 \times F(\lambda)$ plot offers the CWW predicted stellar spectrum. The mean is in *red*; the ± 1-σ bounds flank the mean in *green*; and the six MSX absolute photometric measurements ("MSX global calibration") are in *blue* and indicate both the 1-σ flux uncertainties of the SPIRIT III data and the filters' bandwidths. The observed flatness of these spectra longward of 10–15 μm indicates that cool giants have essentially attained the Rayleigh–Jeans domain by these wavelengths

SPIRIT III data and the filters' bandwidths. The observed flatness of these spectra longward of 10–15 μm indicates that cool giants have essentially attained the Rayleigh–Jeans domain by these wavelengths.

These absolute validations of the CWW calibration basis represent a substantial effort by the AFRL group and I recommend the Price et al. (2004) article as a fascinating description of physics elegantly analyzed. MSX links the optical and MIR regimes. The importance of the MSX validations of these primary and secondary calibrators goes far beyond these individual stars because of the validations of the tertiary CWW stellar network of stars based on the concept of stellar (low-resolution) templates (Cohen et al. 1999). The bright primary and/or secondary calibrators underpinned COBE/DIRBE, airborne MIR spectrometers on NASA's Kuiper Airborne Observatory (the KAO), and a wide variety of ground-based instruments. But the subsequent networks of faint tertiaries of K0–M0III stars provided the absolute calibrators for the joint Japanese–US, IR Telescope in Space (the IRTS), ESA's ISO, NASA's 2MASS and Spitzer, Japan's AKARI, and, the latest NASA MIR surveyor, the Widefield IR Survey Explorer (WISE).

6 Stellar Calibration Networks

Figure 2 shows two very different all-sky networks of absolute stellar calibrators in this Aitoff projection in equatorial coordinates. The 614 red crosses are an expanded version of the Cohen et al. (1999) K and M giant network. These represent

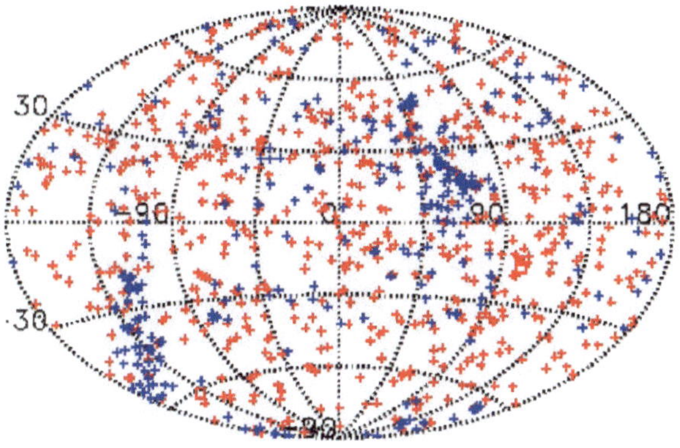

Fig. 2 Two all-sky networks of absolute stellar calibrators in an Aitoff projection in equatorial coordinates. The *red crosses* are an expanded version of the Cohen et al. (1999) K and M giant network. These represent the MIR-brightest normal cool giants in the sky from an IRAS 12-μm flux density of almost 800 Jy to 1 Jy. The *blue crosses* indicate the stars drawn from the Spitzer FEPS Legacy. See text for details

the MIR-brightest normal cool giants in the sky from an IRAS 12-μm flux density of almost 800 Jy to 1 Jy. The blue crosses indicate the 282 stars drawn from the Spitzer FEPS Legacy (The Formation and Evolution of Planetary Systems: Placing Our Solar System in Context: Meyer et al. 2004). The target list for FEPS consisted of 327 roughly solar-type stars (types F through early K). The goal was to seek evidence of dusty debris over a stellar age range of roughly 3 Myr to 10 Gyr. Forty five stars showed departures from their predicted photospheric energy distributions that were interpreted as thermal emission from dust grains. The photospheres of the remaining 282 stars are well represented by Kurucz synthetic spectra on the basis of the assembled optical and IR photometry and IR spectra, much of it based on Spitzer observations. Therefore, these F–K dwarfs are ideal for IR calibrators and can augment the cool star network described above. The template technique was adapted to the creation of far fainter calibrators than had been generated previously, in order to provide Spitzer's IR Array Camera (IRAC) with faint primary and secondary calibration stars (Cohen et al. 2003). The 8-μm dynamic range in the IRAC NEP prime calibrator suite alone is 10,000. Subsequently, Cohen developed custom networks for other Spitzer Legacies such as the 238 calibrators made to support the SAGE study of the Large Magellanic Cloud (Meixner et al. 2006, see their Table 3 for calibrators used in the 3–24-μm range). Such stellar networks ensure a homogeneous calibration over the many Spitzer/IRAC campaigns in which a given Legacy obtains its data. Widespread usage of these networks also enables transparent mingling of data between Spitzer Legacy projects, and indeed between all missions and instruments that use the CWW basis for their absolute calibration. Combining all these bright and faint calibrators that mostly consist of A0–5V and K0–M0III stars, there are now almost 2,000 absolute MIR standard stars, all self-consistent. The MIR dynamic range of this ensemble currently exceeds two million (from α Boo, mag ~ -3 to the faintest IRAC NEP fiducial, mag $\sim +13$). The faintest template calibrator constructed so far is an A0V star predicted to have an IR mag of 19.5 (Cohen 2007). I am still waiting for a volunteer to commit the observing time to test this prediction either in the NIR or the MIR. This faintness, of course, is not the limit to the method. Any star can be converted into an absolute calibrator provided that the star is consistent with the CWW context, and one has secured accurate optical and IR photometry, and an optical spectrum adequate to classify its type and to select a suitable template for that type. Templates may be empirical (CWW) which is more appropriate for cool giants with their plethora of molecular lines that still pose problems for modelists and at high-resolution, or model-atmosphere generated synthetic spectra, e.g., of A dwarfs.

7 Validation of Radiometric Diameters

There is, of course, no better assessment than an independent, absolute, NIST-traceable calibration. However, it is always reassuring to receive validation of any approach to calibration from another branch of the astronomical community. In order to establish their Sirius spectrum, CWW derived a radiometric angular diameter for this star from IR photometry with respect to the Kurucz photospheric spectrum of Vega. The CWW diameter for Sirius was 6.04 ± 0.05 mas. There have been several interferometric measurements of the diameter of Sirius. The earliest yielded 5.89 ± 0.16 mas, limb darkened (LD); (Hanbury Brown et al. 1974: intensity IFR); the next was 5.93 ± 0.08 mas LD (Tango and Davis 1986: amplitude IFR): and the most recent, 6.039 ± 0.019 mas LD, by Kervella et al. (2003: using the VINCI instrument on the VLTI). The latter authors comment on their remarkable agreement with the CWW diameter from 1992. In a wider context, Cohen et al. (1999: their Fig. 8) compared their radiometric diameters for 21 cool giants which had either limb darkened measurements of their interferometric angular diameters or lunar occultation diameters. The correlation was excellent, yielding a slope of 1.013 ± 0.008 and an intercept of 0.035 ± 0.073. Subsequent independent analysis of the 412 published stars calibrated by Cohen et al. (1999) was undertaken by Bordé et al. (2002) who were looking for stars of known angular diameter to build a catalog of long-baseline interferometric calibrators. They accepted 374 of these 412 stars for this purpose after their own examination of the methodology of deriving stellar radiometric diameters.

8 Vega: Unsuitable as a Calibrator?

Vega is now known to be a pole-on rapidly rotating star (e.g., Gulliver et al. 1994) and there has been much discussion about the distribution of material close to this star which affects its NIR emission (e.g., Aufdenberg et al. 2006). The variation in effective temperature from pole to equator strictly invalidates synthetic spectra

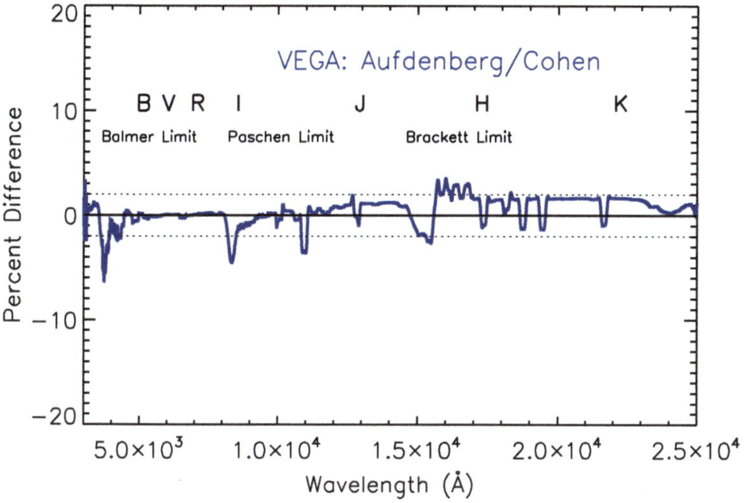

Fig. 3 Differences between the 0.5 and 2.5 μm emergent flux computed from a realistic model for Vega, that accounts for the highly non-uniform distribution of surface flux described by Aufdenberg and colleagues, with that for the 1991 Kurucz spectrum of Vega used by CWW and validated in the NIR and MIR by MSX. The difference is in the sense Aufdenberg minus Kurucz, and the two *dotted lines* indicate ±2%

derived from previous conventional stellar atmosphere models of this star but at what level of predictability do these effects cause old and new models to diverge? Holberg et al. (2008) argue that the absolute optical calibration of Vega is good at the 1% level, despite the physical complexities of the real, as opposed to the ideal, stellar photosphere. Problems arise when extending this calibration into the NIR where discrepancies of the order of 2% are found.

I am indebted to Jay Holberg for Fig. 3 in which he differences the 0.5–2.5 μm emergent flux computed from a realistic model for Vega, that accounts for the highly non-uniform distribution of surface flux described by Aufdenberg and colleagues, with that for the 1991 Kurucz spectrum of Vega used by CWW and validated in the NIR and MIR by MSX. The uncertainties in the computation based on the Aufdenberg model are at least 2%, whereas CWW assign 1.45% to the uncertainties in their Kurucz spectrum. The two continua are very similar. The spectral differences are mostly within the 2% level. Where there exist greater excursions, these usually correspond to H series' limits. At such wavelengths the discrepancies between models are amplified even by slight differences in spectral resolution or in the treatment of the physical processes. Consequently, there is not a large discrepancy between calibration based on an old model and on a new, more detailed, Vega model. But if one seeks to attain levels of accuracy higher than the ±1% level and wishes to extrapolate well outside the visible range then there are reasons for caution.

Another reason to question the suitability of Vega as a primary standard derives from the long history of variability of this star (see Engelke et al. 2010 for a detailed discussion). During the 1930s several observers noted photometric variations in Vega (e.g., Fath 1935). Later, Fernie (1981) reported variations at the same level as seen by Fath, about 0.02 mag. Even more recently, (Engelke et al. 2010) have gathered 90 measurements from the Hipparcos archives that show clear variability in Vega at the 6% level between 1990 and 1993. One can understand the difficulties in 1930 in discerning variability at the 2% level, yet, from the ground, even now, it cannot be easy, to judge by a comment by Alekseeva et al. (1996) in an article describing The Pulkovo Spectrophotometric Catalog of Bright Stars: "Vega.. has been used as the primary standard. Although some authors suspected this star to be variable, the latest observations of high accuracy at the Vilnius Observatory do not find any changes of its brightness."

MSX's adoption of a specific photospheric spectrum of Vega to define zero magnitude rather than the real star and the dependence on Sirius as the primary (also governed by the desire for higher signal-to-noise) have proven to be excellent choices.

9 Replacing the Real Vega as the Zero Magnitude Fiducial?

If one accepts that Vega is no longer a suitable fiducial star what is the alternative? Engelke et al. (2010) have suggested that an amalgam of the A0V star 109 Vir and Sirius would provide a robust empirical reference. 109 Vir would serve from 0.33 to 0.90 μm, and Sirius for wavelengths beyond these. These authors show that 109 Vir displays no variations like those of Vega in Hipparcos data over the same period of time. Although not as frequently measured with respect to a standard lamp as is Vega, 109 Vir was observed in this fashion by Tug et al. (1977), who state the absolute uncertainties in this star's energy distribution to be $\pm 1 - 2\%$ in the red and blue respectively, barring systematic errors. Engelke et al. compare their proposed new zero points with those of Rieke (2008), find consistency with those based on solar analog stars, and estimate the uncertainties in their new zero points to be of order 1%.

10 A Concern About Model Atmospheres

I do have a major concern about the reliance on modeled synthetic spectra of photospheres and that is their changeablity. Theoreticians who model stars are always seeking to upgrade the realism of their physics, to include another billion molecular lines as these giant databases (like that of water vapor) become available for inclusion. Last year's model is not the same as this year's although, apparently, there are fewer modelists now than in the past, even for the more readily modeled A-dwarfs. Bohlin and Cohen (2008), in quest of achieving JWST calibrators with 1% or higher precision, found differences between model spectra (and in the treatment of modeled physics) from old and more recent grids of models that significantly exceed the 1%

goal. Modelists should always provide the errors, preferably wavelength-dependent, that are associated with the shapes of their synthetic spectra. Based on comparisons between observed A-dwarf spectra and those modeled by different groups, the CWW context has always assigned 5% a absolute uncertainty, independent of wavelength, to any photospheric synthetic spectrum. Changes in spectra between releases of model grids can reveal differences at this level and Bohlin and Cohen verified that 5% is still a realistic uncertainty, particularly in the IR regime, longward of 2.5 μm. Their conclusion is in accord with that of Holberg et al. (2008), which was based on Vega's spectrum.

11 Another New Approach to Calibration

Finally, ACCESS is an approved rocket-borne payload with a NIST-traceable ground calibration (Kaiser et al. 2010). Its goal is to transfer a 1% accurate ground calibration to a set of standard stars in the 0.35–1.7 μm region at a spectral resolution of R = 500 to support dark energy missions based on SNe.

Acknowledgments Support for my calibration work came from AFRL for MSX; SAO for Spitzer/IRAC and NASA via UCLA for WISE. I thank Dr. S. D. Price for an early draft of the Engelke, C.W., Price, S.D., Kraemer, K.E., paper.

References

Alekseeva, G.A., et al. (1996). *Baltic Astronomy*, **5**, 603
Aufdenberg, J.P., et al. (2006). *Astrophysical Journal*, **645**, 664
Blackwell, D.E. (1991). Private Communication
Bless, R.C., Code, A.D., & Fairchild, E.T. (1976). *Astrophysical Journal*, **203**, 410
Bohlin, R.C. (1996). *Astronomical Journal*, **111**, 1743
Bohlin, R.C., & Cohen, M. (2008). *Astronomical Journal*, **136**, 1171
Bohlin, R.C., Colina, L., & Finley, D.S. (1995). *Astronomical Journal*, **110**, 1316
Bohlin, R.C., Harris, A.W., Holm, A.V., & Gry, C. (1990). *Astrophysical Journal, Supplement*, **73**, 413
Bohlin, R.C., Sparks, W.M., Holm, A.V., Savage, B.D., & Snijders, M.A.J. (1980). *Astronomy and Astrophysics*, **85**, 1
Bordé, P., Coude du Foresto, V., Chagnon, G., & Perrin, G. (2002). *Astronomy and Astrophysics*, **393**, 183
Brune, W.H., Mount, G.H., & Feldman, P.D. (1979). *Astrophysical Journal*, **227**, 884
Carnochan, D.J. (1982). *Monthly Notices of the Royal Astronomical Society*, **201**, 1139
Carruthers, G.R., Heckathorn, H.M., & Opal, C.B. (1981). *Astrophysical Journal*, **243**, 855
Cohen, M. (2007). "Networks of Absolute Calibration Stars for SST, AKARI, and WISE," in C. Sterken (ed.), *The Future of Photometric, Spectrophotometric and Polarimetric Standardization*, (San Francisco: ASP Conference Series), **364**, 333
Cohen, M., Megeath, T.G., Hammersley, P.L., Martin-Luis, F., & Stauffer, J. (2003). *Astronomical Journal*, **125**, 2645
Cohen, M., Walker, R.G, Barlow, M.J., & Deacon, J.R. (1992). *Astronomical Journal*, **104**, 1650

Cohen, M., Walker, R.G., & Witteborn, F.C. (1992). *Astronomical Journal*, **104**, 2030 (CWW)

Cohen, M., Walker, R.G., Carter, B., Hammersley, P.L., Kidger, M.R., & Noguchi, K. (1999). *Astronomical Journal*, **117**, 1864

Dreiling, L.A., & Bell, R.A. (1980). *Astrophysical Journal*, **241**, 736

Engelke, C.W., Price, S.D., & Kraemer, K.E. (2010). *Astronomical Journal*, **140**, 1919

Fath, E.A. (1935). *Lick Observatory Bulletin*, **17**, 115

Fernie, J.D. (1981). *Publications of the Astronomical Society of the Pacific*, **93**, 333

Gulliver, A.F., Hill, G., & Adelman, S.J. (1994). *Astrophysical Journal*, **429**, 81

Hanbury Brown, R., Davis, J., & Allen, L.R. (1974). *Monthly Notices of the Royal Astronomical Society*, **167**, 121

Hayes, D.S. (1985). In D.S. Hayes, L.E. Pasinetti, & A.G.D. Philip (eds.), *Calibration of Fundamental Stellar Quantities*, (Dordrecht: Reidel), p. 225

Henry, R.C., et al. (1975). *Astrophysical Journal*, **201**, 613

Holberg, J.B., & Bergeron, P. (2006). *Astronomical Journal*, **132**, 1221

Holberg, J.B., & Polidan, R.S. (1985). Proceedings of the International Astronomical Union Symposium 111, (Dordrecht: Reidel), p. 479

Holberg, J.B., Bergeron, P., & Gianninas, A. (2008). *Astronomical Journal*, **135**, 1239

Holberg, J.B., Forrester, W.T., Shemansky, D.E., & Barry, D.C. (1982). *Astrophysical Journal*, **257**, 656

Kaiser, M.E., et al. (2010). *Bulletin of the American Astronomical Society*, **42**, 404

Kervella, P., et al. (2003). *Astronomy and Astrophysics*, **408**, 681

Kurucz, R.L. (1979). *Astrophysical Journal, Supplement*, **40**, 1

Kurucz, R.L. (1991). private communication cited in Cohen et al. (1992)

Lampton, M., Margon, B., Paresce, F., Stern, R., & Bowyer, S. (1976). *Astrophysical Journal*, **203**, L71

Megessier, C. (1995). *Astronomy and Astrophysics*, **296**, 771

Meixner, M., et al. (2006). *Astronomical Journal*, **132**, 2268

Meyer, M.R., & 36 coauthors (2004). *Astrophysical Journal, Supplement*, **154**, 422

Mountain, C.M., Selby, M.J., Leggett, S.K., Blackwell, D.E., & Petford, A.D. (1985). *Astronomy and Astrophysics*, **151**, 399

Price, S.D., et al. (2004). *Astronomical Journal*, **128**, 889

Rieke, G.H. (2008). *Astronomical Journal*, **135**, 2245

Selby, M.J., Blackwell, D.E., Petford, A.D., & Shallis, M.J. (1980). *Monthly Notices of the Royal Astronomical Society*, **193**, 111

Selby, M.J., Mountain, C.M., Blackwell, D.E., Petford, A.D., & Leggett, S.K. (1983). *Monthly Notices of the Royal Astronomical Society*, **203**, 795

Sing, D., Holberg, J.B., & Dupuis, J. (2002). *Astronomical Society of the Pacific Conference Series*, **264**, 57

Strongylis, G.J., & Bohlin, R.C. (1979). *Publications of the Astronomical Society of the Pacific*, **91**, 205

Tango, J., & Davis, W.J. (1986). *Nature*, **323**, 234

Tremblay, P.-E., & Bergeron, P. (2007). *Astrophysical Journal*, **657**, 1013

Tug, H., White, N.M., & Lockwood, G.W. (1977). *Astronomy and Astrophysics*, **61**, 679

Optical Region Spectrophotometry: Past and Present

Saul J. Adelman

1 Introduction

I first became acquainted with the technique of optical region spectrophotometry with rotating grating photomultiplier tube scanners about 1968 when I was a graduate student at the California Institute of Technology. At this time these flux measurements were an important part of stellar astrophysics. Oke (1965) provides an excellent review of the field at that time. The legacy data produced by scanners are still of considerable importance. For the last 40 years, I have been a spectrophotometric observer, an analyzer of the resulting flux data, a compiler of spectrophotometric fluxes, a sponsor of a meeting on spectrophotometry, an advocate for a next generation of instrumentation, and finally the somewhat frustrated builder of a nearly completed automated spectrophotometric telescope system.

2 History

Spectrophotometric fluxes are spectroscopic data obtained with photometric observation techniques. Typically their resolutions are similar to those of classical spectral calibration or objective prism spectroscopy and have signal-to-noise values greater than 100:1. These observations should be attempted only when the sky conditions are photometric. One needs standard stars absolutely calibrated by appropriate techniques. These stars should be observed on nights when one measures other targets also. The secondary standards are usually used to both measure the nightly extinction as well as to provide a way to convert the instrumental system values to the absolute system. The proceedings of *New Directions in Spectrophotometry* (ed. by Philip et al. 1988) is a testament to how astronomers who were interested in spectrophotometry thought about the subject 20 years ago. Of special interest is the paper

S.J. Adelman (✉)
Department of Physics, The Citadel, 171 Moultrie Street, Charleston, SC 29409, USA
e-mail: adelmans@citadel.edu

E.F. Milone and C. Sterken (eds.), *Astronomical Photometry: Past, Present, and Future*, Astrophysics and Space Science Library 373, DOI 10.1007/978-1-4419-8050-2_9,

by Benjamin Taylor (1988) on *The History and Legacy of Photomultiplier Scanners*. Astronomers were interested in topics including the spectrophotometry of normal and peculiar stars of Population I and II, the broad continuum features of the chemically peculiar (CP) stars of the Upper Main Sequence, the use of spectrophotometry to determine the parameters of binary stars with composite spectra, the determination of stellar effective temperatures and surface gravities, the determination and the removal of the effects of interstellar reddening, spectrophotometric standards in the optical region and in the ultraviolet (especially for IUE and the HST), the spectrum synthesis of Population II systems, the synthesis of photometric data from spectrophotometry, and the archival and the dissemination of spectrophotometric data.

I discuss investigations done primarily in the United States, the UK, and Australia (the West). There was also a substantial Soviet spectrophotometry establishment. But I found its results rarely agreed well with my own and other Western results due to larger errors. Thus I usually did not use them.

Astronomers by the late 1930s realized the importance of measuring the absolute energy distributions of stars. Investigations working towards this goal were conducted in Germany, the United Kingdom, the United States, France, and other countries. After World War II, the 1P21 photomultiplier tube became available and after some time astronomers began to use it. The first uses were for photometry with spectrophotometry a later use. For photometry one could use a filter wheel to make measurements through a number of filters. For spectrophotometry one could not follow that route as measurements were made at many more wavelength settings. We are making sequential measurements between which seeing fluctuations can cause important differential and mean changes. Thus the minimum exposure time usually selected was 10 s. Two kinds of instruments were developed:

1. In the photomultiplier tube spectrographs, narrow spectrograph slits kept the resolution high. Seeing was dealt with using scan and monitor tubes (Code and Liller 1962).
2. In the rotating grating or flux-curve scanners all the stellar light passed through an aperture much larger than the stellar image. The stellar image limited the resolution to a few angstrom (Oke 1965).

The Cambridge observers became proficient with photomultiplier spectrographs. But far more scanners were built and used regularly. Some examples are the Wampler (1966) scanner at Lick Observatory, the Palomar Observatory scanner, the Mt. Wilson Observatory scanner, the Harvard College Observatory scanner used at Kitt Peak National Observatory, the University of Wisconsin scanner, and the Indiana University scanner. There are also 5-m Hale telescope measurements, based largely on rotating grating scanner calibrations, with now retired instruments: the multichannel spectrophotometer (Oke and Schild 1983; Gunn and Stryker 1983) and the Double Spectrograph (Oke 1990, who provided standards for the Hubble Space Telescope) and satellite measurements of the optical ultraviolet (e.g., Code and Meade 1979).

After these instruments, none capable of obtaining data similar or better than the rotating grating scanners were built. When I discussed the capabilities of some newer instruments built in the United States using different designs with the builders and their advocates, I was given incorrect information about their stellar spectrophotometric capabilities. The reality was their replacements were not intended for high quality stellar observations and lacked the required accuracy and precision. Without new modern instruments, the field has become marginally active. The Wampler scanner and the Harvard College Observatory scanner continued to be used until a few years ago. Despite what has happened over the last 20 years the technique of optical spectrophotometry can be revived with an improved absolute calibration and a single properly designed instrument. It is desirable to have one in both the Northern and the Southern Hemispheres. One modern instrument should be able in a year with many photometric nights obtain sufficient higher quality data to replace most of the published fluxes.

Breger (1976a), Ardeberg and Virdefors (1980), and Adelman et al. (1989) compiled observations. Breger selected fluxes from the literature whose observations were obtained at continuum and near continuum wavelengths so these values could be used to derive stellar temperatures and surface gravities as well as to calibrate wide and intermediate band photometric systems. Ardeberg and Virdefors followed his lead while the Adelman et al. paper is a compilation of the values my collaborators and I obtained.

3 Scanners

Despite their success, the scanners had problems which were known to spectrophotometrists. Their resultant stellar fluxes may contain systematic wavelength-dependent errors due to those in the absolute calibration, extinction, bandpass centering, and scattered light in the instrument. Such data typically consists of 15–20 values covering $\lambda\lambda$3,400–7,100 with 20–50 Å wide bandpasses usually of spectral regions with minimal line blanketing. The scanners I used were built before LTE H-line blanketed and line blanketed model atmospheres had been calculated. So the choices of bandpasses are understandable. It was also an epoch during which astronomers deblanketed spectrophotometric fluxes for comparison with the model atmospheres of the day. But the need to use filters to separate the first and second order spectra lead to discontinuities in the resulting fluxes, as often to observe a single order, one had to use at least two different filters. In some of them, one entered the program to be followed for observing each star (the Harvard College Observatory scanner) whereas for others one had to change the wavelengths and filters manually (the Palomar Observatory scanner). Data obtained with two scanners could show differences, i.e., Harvard College scanner vs. Palomar Observatory scanner in the optical ultraviolet. The Mt. Wilson Observatory scanner (see Fig. 1) had a red leak which meant that measurements made at some central wavelengths did not yield consistent results.

Due to time constraints the extinction often was based on mean observatory values with errors rarely better than 1%. If one observed twice at 20 wavelengths, took 10 s exposures, and allowed 4 s between measurements, then a bright star

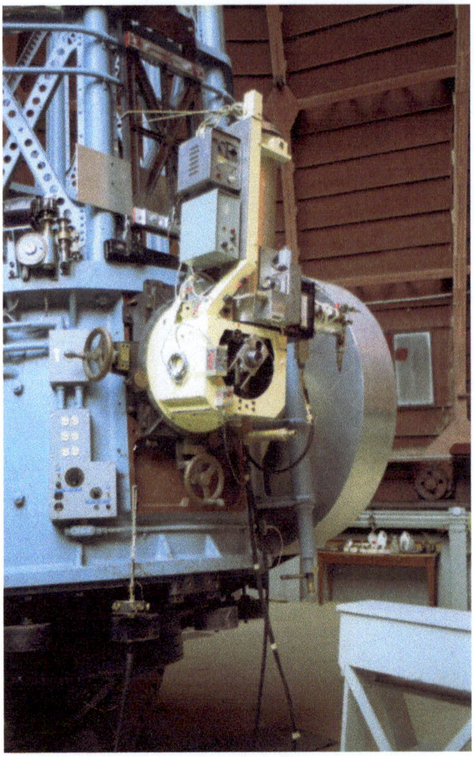

Fig. 1 The Mt. Wilson 60″ Scanner

observation at KPNO or Palomar took about 10 min. Typically only 6–9 standards were observed on full photometric nights. At most observatories this is not a really satisfactory way to define the extinction. Astronomers who perform absolute photometry advocate that one should observe 6–10 standards per hour. But for spectrophotometry the time per observation with a scanner was much greater than for filter photometry; and so far fewer standards were obtained with spectrophotometry. There was always a competition between taking sufficient standards and spending time to get data of one's program stars.

With optical region grating scanner (and ultraviolet flux) data and Balmer line profiles, astronomers derived reasonably good effective temperatures and surface gravities of normal single slowly rotating B, A, and F stars using LTE line blanketed model atmospheres especially ATLAS9 (Kurucz 1993). These have been expressed in terms of filter photometric indices of various systems. For stars with significant non-solar compositions, such calibrations are not necessarily accurate, as metallicity, microturbulence, macroturbulence, and/or magnetic fields affect the stellar fluxes in subtle, yet measurable ways (Adelman and Rayle 2000). For stellar astrophysics, at the heart of our understanding the history and evolution of galaxies, accurate spectrophotometry is critical for future advancements. Spectrophotometry can also be an important technique for the study of solar system objects, nebulae, star clusters, and galaxies. *New Directions In Spectrophotometry* (eds. Philip et al. 1988) discusses additional uses.

4 Calibration

Oke (1965) discusses the absolute calibration of standard stars and the techniques then used. It is necessary to have spectrophotometric stars distributed over the sky which can be used to calibrate the scanner in an absolute way. Two basic problems are to calibrate some primary standard (often Vega) in terms of a known black-body source usually on the Earth's surface and to compare the proposed standard stars with the primary standard. One usually chose enough wavelengths to provide the absolute calibration completely or measured at sufficiently close wavelengths so that the interpolation of the absolute energy distribution can be made accurately.

Breger (1976a) used studies by Bahner (1963), Oke (1964), Hayes (1970) and Schild et al. (1971) to derived his standard system. But when the Hayes and Latham (1975) calibration of Vega was published, he converted his catalog values to their system as it was more fundamentally derived. Taylor worked to improve the system of secondary stars. His 1984 paper provided values for additional secondary standards as well as extended the results for those stars used for this purpose. He used primarily the values of Breger (1976b) for $\leq\lambda$ 6,058 and longward of that from his own work presented in this paper and his 1979 paper and that of Cochran (1981). Taylor (2007) continued this process.

All ground-based stellar photometry is relative in the sense that the program stars are compared with the standard stars. When the fluxes are extrapolated to outside the Earth's atmosphere, the extinction errors tend to cancel – but not in the case of stellar flux calibrations because the stars are compared with a nearby terrestrial standard. Here errors in the extinction coefficients are multiplied by the mean air mass instead of the difference in mean air mass at which the stars were observed. So stellar-flux calibration is much more sensitive to extinction errors. There are errors both in the vertical extinction to the star and the horizontal extinction to the standard source (Hayes et al. 1975). Hayes and Latham (1975) reviewed both the Lick (Hayes 1970) and the Palomar (Oke and Schild 1970) absolute calibrations of Vega and found that the atmospheric extinction was treated incorrectly. They modeled the extinction in the Earth's atmosphere and used their results to calculate corrections to both calibrations.

Experience with Vega and with photometric attempts to establish standards for many systems show that using one star as the standard is not an optimal choice. Because in the past, one could observe in only one bandpass at a time, it was a pragmatic necessity. Now with detectors that can record multiple passbands, the multiplexing ability should lead to a grid of well calibrated standards calculated against one another and absolute standard lamps.

Gulliver et al. (1994) discovered that Vega was a flat rotating star seen nearly pole on. Thus it is difficult to model, but is acceptable for use as a spectrophotometric standard. Its predicted fluxes, however, cannot be calculated from a single model atmosphere with a given effective temperature and surface temperature for an extended wavelength range.

5 ACCESS

Attempts have been made to perform absolute calibrations from space to remove the effects of the Earth's atmosphere. Once ultraviolet fluxes were measured it was found that the derived fluxes from certain white dwarfs did not match the theoretical power law distribution. So indirect means were used to get a calibration. Bohlin (2007) describes the HST CALSPEC Stellar Standards which he claims have a 1% accuracy in absolute flux. This standard star network is based on observations of the pure hydrogen white dwarf stars G 191B2B, GD 153, and GD 71. For these stars, the models were calculated with TLUSTY (Hubeny and Lanz 1995). They are normalized to that of Vega. For comparison Cohen (2007) and in the present volume, examines absolute calibration in the infrared region. Although these HST Standards are most likely better than the Hayes and Latham (1975) calibration of Vega for absolute work, I am somewhat uneasy that they do not depend on a more direct comparison between primary standard star fluxes and laboratory irradiance standards.

Fortunately the ACCESS (absolute color calibration experiment for standard stars) project (Kaiser et al. 2008) is following this path. It is a series of rocket-borne sub-orbital missions and ground-based experiments to enable absolute flux measurements for a limited number of primary stars. ACCESS uses a Dall-Kirkham cassegrain telescope with aluminum and fused silica overcoated Zerodur mirrors. The spectrograph is a low-order echelle with a cooled, substrate removed, HgCdTe detector. Three orders are used: first order ($\lambda\lambda$9,000–19,000), second order ($\lambda\lambda$ 4,500–9,500), and third order ($\lambda\lambda$3,000–6,333). Two optical elements, a concave low ruling density diffraction grating and a prism with spherically figured surfaces placed in the converging beam, produce a separation between orders of about 1 mm on the detector. ACCESS utilizes a ground-based calibration and performance monitoring program that is designed to transfer the National Institute of Standards and Technology absolute laboratory standards to stars. The resulting precision and calibration accuracy should be 1% for $\lambda\lambda$3,500–17,000. The spectral resolving power $R = 500$. This is a significant improvement in both absolute and relative astrophysical fluxes, especially for near-infrared wavelengths.

To minimize calibration uncertainties, this project: (1) judiciously selects standard stars, (2) makes observations above the Earth's atmosphere, (3) uses a single optical part and detector, (4) establishes an a priori error budget, (5) utilizes on-board monitoring of instrumental performance, and (6) fits stellar model atmospheres to the data to search for discrepancies and to confirm performance. The primary stars to be calibrated are Vega, Sirius, the $V = 8.4$ mag. Spitzer/IRAC standard HD 37725, and the $V = 9.5$ mag Sloan Digital Sky Survey standard BD+ 17 4708. The observing time above the atmosphere per flight is limited to approximately 400 s. Two flights are needed for each of two observing fields (Vega and BD+17 4708; Sirius and HD 37725) to check repeatability to $<$1%.

6 The ASTRA Spectrophotometric Telescope

I have been working on a new generation spectrophotometer on and off for about 20 years. My two primary professional collaborators are Austin Gulliver (Brandon University, CA) and Barry Smalley (Keele University, UK). We have spoken to a number of other astronomers who would like to work with us in reducing the substantial amount of data anticipated. In its latest reincarnation, my associates and I named it the ASTRA (Automated Spectrophotometric Telescope Research Associates) Spectrophotometer (Adelman et al. 2007; Smalley et al. 2007). As I have observed at the Dominion Astrophysical Observatory since 1984, it was natural to speak with E. Harvey Richardson. He suggested I consult John Pazder, his then graduate student, who became the project's optical designer. Frank Younger was the mechanical designer. To fill the instrumental void so that astronomers could again measure high quality optical region stellar fluxes, we designed and built the ASTRA spectrophotometer for use with a 0.5-m automated telescope located at the Fairborn Observatory, Washington Camp, Arizona, where Lou Boyd is the director (Figs. 2 and 3). It is the first such instrument built to measure simultaneously in detail nearly complete optical region energy distributions of stars with high accuracy

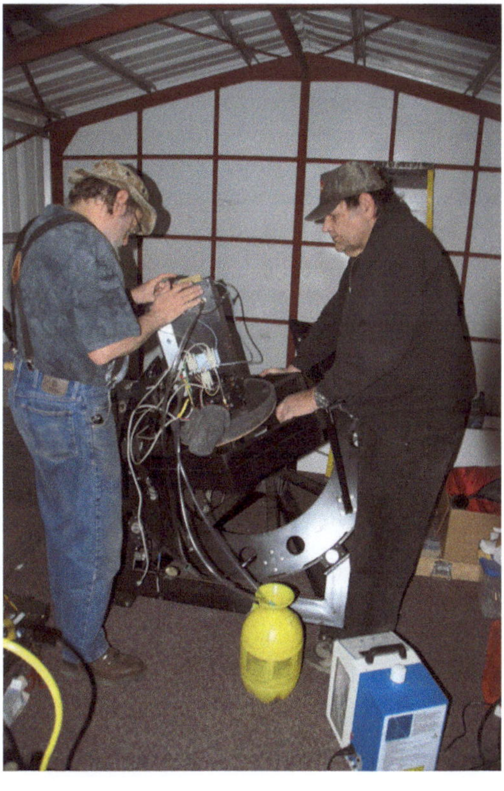

Fig. 2 Louis Boyd (*left*) and Frank Younger (*right*) with the spectrophotometer attached to what will be the back of the telescope

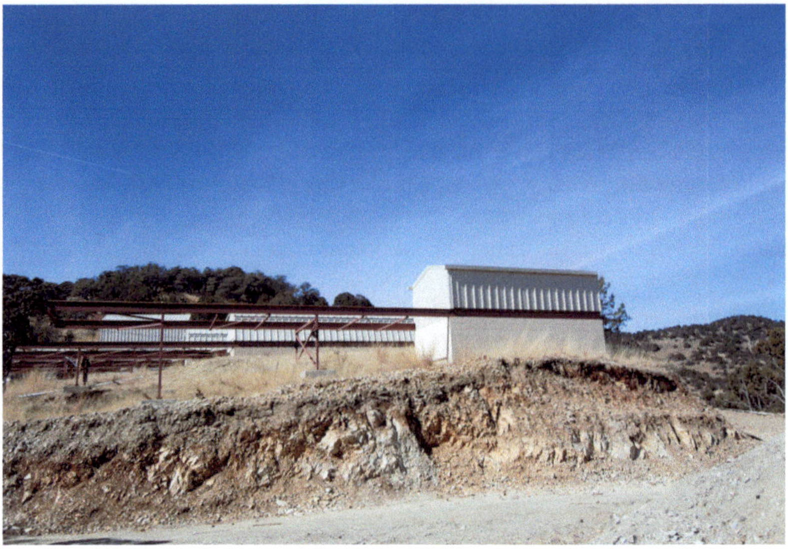

Fig. 3 ASTRA Spectrophotometer building

and precision. Its scientific detector is a back-illuminated CCD. The spectrophotometer is finished, waiting for its telescope to be completed.

Our telescope and observatory choice was made after an extensive study. I have considerable experience as an astronomer using the Four College Automated Photometric Telescope at Fairborn Observatory. A 0.5-m automated telescope was custom built for exclusive use with our spectrophotometer. Automation has the advantages of uniformity of data collection and minimization of slew times. Further, no observer is needed and the costs associated with observing are minimized. We wanted to observe the brightest stars (through neutral density filters). The other major innovation is to use a CCD as a detector and to measure simultaneously both the first and second order spectra. This sets major constraints on the design which uses a prism as a cross-disperser. As the bandpasses in the red are twice as wide as those in the blue, many optical parts have increased blue sensitivities.

Some design criteria for our spectrophotometer follow: They incorporate wisdom from optical designers, photometrists, and observers who used scanners.

1. Minimize cost by using simple optical and mechanical designs
2. Minimize the number of optical surfaces to reduce light loss and scattering
3. Mount the optical components so that they will not drift out of alignment
4. Make the spectrophotometer compact to reduce moments on the telescope
5. The average seeing at Fairborn Observatory is 1.5–2.0 arcsec
6. The 2-pixel resolution is less than 15 Å
7. The science CCD must have a high quantum efficiency for $\lambda\lambda$3,200–9,500 especially shortward of the Balmer jump

8. The spectrograph must have good sensitivity between $\lambda\lambda$3,300–9,000 with short- and long-wavelength extensions desirable
9. Use a prism cross-disperser to obtain the entire desired spectral range with one exposure
10. Use a square projected aperture for accurate sky subtraction
11. Use the zeroth order light from the spectrograph to guide the telescope
12. Widen the spectrum to at least 5 pixels to be able to remove properly any cosmic ray signatures
13. Temperature-control the spectrograph for stability. (The science CCD was purchased from Apogee with a temperature-control system)
14. Use a wide-field CCD camera to center the stellar image on the spectrophotometer entrance aperture after the initial telescope pointing

The spectrophotometer's insulated case is a box, rectangular in cross-section which may be opened to provide access to the instrument. The length overall is roughly 38 cm, greatest width 28 cm, and height 14 cm. The box front is the mounting disc which couples with the back of the telescope, centered on its optical axis. A hole in the disc center holds a baffle, the upper end of which is sealed with a 1-cm thick fused silica window. The optical plate is a 1.25-cm thick aluminum; the remainder of the case structure was made from 0.95-cm aluminum. The mass is about 20.5 kg (45 pounds). Thermal stability is accomplished by using 4°C water to cool the optical plate and covering the entire spectrophotometer with 1.25-cm maritime insulation. The instrument is appropriately baffled with black flocking to minimize scattering problems.

The optical parts are epoxied into solid mounts. With our slitless spectrophotometer, focusing is done with the telescope. The optics placement on the optical plate used an illuminated end of a 100 μ optical fiber placed at the theoretical focal position inside the box. The grating rotation was set during the alignment. Both a He/Ne laser and an illuminated 100 μ fiber were used to align the optics and set the grating rotation.

A prismatic cross disperser provides sufficient order separation for the spectrophotometer to cover $\lambda\lambda$3,000–10,000 in a single exposure. The main dispersion element is a 300 gr/mm grating with a λ8,600 blaze. From diffraction grating efficiency data, the optimal order coverage is $\lambda\lambda$5,500–10,000 in the first and $\lambda\lambda$3,000–6,000 in the second order. A 500 Å overlap of the orders allows the data quality to be checked.

A 1.0 arcsec object is 2 pixels wide at the image, which is the Nyquist frequency of the 26 square micron CCD pixels. The optical performance of the spectrophotometer at 80% encircled energy is better than 17 μ (50% in 8 μ) over the whole spectral range for a point source object. A 1.0 arcsec image of the star with a width of 30 μ for a perfect camera, will have an image size, at worst, of 35 μ. As the smallest bandpass is 2 pixels wide, the resolution is 14 Å in the first and 7 Å in the second order.

To preserve the resolution set by the stellar image in this slitless design and to find cosmic ray hits, the spectrum is widened to 5 or 6 pixels by mechanical rocking of the telescope. The separation of the two orders is sufficient so that during the rocking, the sky exposures of each order do not overlap. The 30×30 arcsec projected hole in the stellar acquisition mirror used to acquire the star acts as a field stop. As the CCD read noise is about 8 electrons per pixel, widening the spectrum does not significantly degrade the S/N ratio.

The guide and centering stellar acquisition camera optics are both standard achromatic doublets. For the guide camera the image scale is 2 pixels for a 1.0 arcsec disc, and for the stellar acquisition camera it is 3 pixels for a 1.0 arcsec disc.

Our first two major planned projects are the revision and extension of secondary standards and sample fluxes of Population I and II stars. Two important auxiliary projects are the comparison of fluxes with model atmospheres and synthetic colors and line indices from spectrophotometry.

Acknowledgments I appreciate my discussions on spectrophotometry with John B. Oke, Donald S. Hayes, Benjamin Taylor, Ralph Bohlin, and Mary Beth Kaiser. My partners in the ASTRA Spectrophotometer Project are Austin F. Gulliver, Barry Smalley, John S. Pazder, Frank Younger, Louis J. Boyd, Donald Epand, and Thomas Younger. This work was supported by NSF Grant AST-0115612.

References

Adelman, S. J., Gulliver, A. F., Smalley, B., et al. (2007). "The ASTRA Spectrophotometer: Design and Overview," in C. Sterken (ed.), *The Future of Photometric, Spectrophotometric, and Polarimatric Standardization*. (San Francisco: ASP) *ASP Conference Series*, Vol. **364**, pp. 255–264

Adelman, S. J., Pyper, D. M., Shore, S. N., et al. (1989). "A Catalog of Stellar Spectrophotometry." *Astronomy and Astrophysics, Supplement*, **91**, 221–223

Adelman, S. J., & Rayle, K. E. (2000). "On the Effective Temperatures, Surface Gravities, and Optical Region Fluxes of the CP Stars," *Astronomy and Astrophysics*, **355**, 308

Ardeberg, A., & Virdefors, B. (1980). "A Catalog of Stellar Spectrophotometric Data." *Astronomy and Astrophysics, Supplement*, **40**, 307–318

Bahner, K. (1963). "Energy Distribution in the Spectra of Early-Type Stars." *Astrophysical Journal* **138**, 1314–1315

Bohlin, R. C. (2007). "HST Stellar Standards with 1% Accuracy in Absolute Flux," in C. Sterken (ed.), *The Future of Photometric, Spectrophotometric, and Polarimatric Standardization*. (San Francisco: ASP) *ASP Conference Series*, Vol. **364**, pp. 315–331

Breger. M. (1976a). "Catalog of Spectrophotometric Scans of Stars." *Astrophysical Journal, Supplement*, **32**, 7–89

Breger, M. (1976b). "Evaluation of Stellar Spectrophotometry." *Astrophysical Journal, Supplement*, **32**, 1–6

Cochran, A. L. (1981). "Spectrophotometry with a Self Scanned Silicon Photodiode Array. II. Secondary Standard Stars." *Astrophysical Journal, Supplement*, **45**, 83–96

Code, A. D. and Liller, W. C. (1962). "Direct Recording of Stellar Spectra," in W. A. Hiltner (ed.), *Asotronomical Techniques*. (Chicago: University of Chicago Press), pp. 281–301.

Code, A. D., & Meade, M. R. (1979). "Ultraviolet Photometry from the Orbiting Astronomical Observatory. XXXII–an Atlas of Ultraviolet Spectra." *Astrophysical Journal, Supplement*, **39**, 195–289

Cohen, M. (2007). "Networks of Absolute Calibration Stars for SST, AKARI, and WISE." In C. Sterken (ed.), *The Future of Photometric, Spectrophotometric, and Polarimatric Standardization*. (San Francisco: ASP) *ASP Conference Series*, Vol. **364**, pp. 333–353

Gulliver, A. F., Hill, G., & Adelman, S. J. (1994). "Vega: A Rapidly Rotating Pole-on Star." *Astrophysical Journal*, **429**, L81

Gunn, J. E., & Stryker, L. L. (1983). "Stellar Spectrophotometric Atlas, Wavelengths from 3130 to 1080." *Astrophysical Journal, Supplement*, **52**, 121–153

Hayes, D. S. (1970). "An Absolute Calibration of the Energy Distribution of Twelve Standard Stars." *Astrophysical Journal*, **159**, 165–176

Hayes, D. S., & Latham, D. W. (1975). "A Rediscussion of the Atmospheric and Absolute Spectral-Energy Distribution of Vega." *Astrophysical Journal*, **197**, 593–601

Hayes, D. S., Latham, D. W., & Hayes, S. H. (1975). "Measurements of the Monochromatic Flux from Vega in the Near-Infrared." *Astrophysical Journal*, **197**, 587–592

Hubeny, I., & Lanz, T. (1995). "Non-LTE Line-Blanketed Model Atmospheres of Hot Stars. I. Hybrid Complete Linearization/Accelerated Lambda Iteration Method." *Astrophysical Journal*, **439**, 875–904

Kaiser, M. E., Kruk, J. W., McCandliss, S. R., et al. (2008). "ACCESS: Enabling an Improved Flux Scale for Astrophysics, in Ground-Based and Airborne Instrumentation for Astronomy II," in I. S. McLean & M. M. Casali (eds.), *Proceedings of the SPIE* (Vol. **7015**, pp. 70145Y-70145Y-14)

Kurucz, R. L. (1993). *Atlas9 Stellar Atmospheres Programs and 2 km/s grid, Kurucz CD-Rom No. 13*. (Cambridge, MA: Smithsonian Astrophysical Observatory)

Oke, J. B. (1964). "Photometric Spectrophotometry of Stars Suitable as Standards." *Astrophysical Journal*, **140**, 689–693

Oke, J. B. (1965). "Absolute Spectral Energy Distributions in Stars." *Annual Review of Astronomy and Astrophysics*, **3**, 23–46

Oke, J. B. (1990). "Faint Spectrophotometric Standards." *Astronomical Journal*, **99**, 1621–1632

Oke, J. B., & Gunn, J. E. (1983). "Secondary Stars for Absolute Spectrophotometry," *Astrophysical Journal*, **266**, 713

Oke, J. B., & Schild, R. E. (1970). "The Absolute Spectral Distribution of Alpha Lyrae." *Astrophysical Journal*, **161**, 1015–1023

Philip, A. G. D., Hayes, D. S., & Adelman, S. J. (eds.) (1988). *New Directions in Spectrophotometry*. (Schnectady, NY: L. Davis Press)

Schild, R., Peterson, D., & Oke, J. B. (1971). "Effective Temperatures of B- and A-type Stars." *Astrophysical Journal*, **166**, 95–108

Smalley, B., Gulliver, A. F., & Adelman, S. J. (2007). "The ASTRA Spectrophotometer: Reduction and Flux Calibration," in C. Sterken (ed.), *The Future of Photometric, Spectrophotometric, and Polarimatric Standardization*. (San Francisco: ASP) *ASP Conference Series*, Vol. **364**, pp. 265–274

Taylor, B. J. (1984). "An Augmented System of Secondary Standards for Bright-Star Spectrophotometry." *Astrophysical Journal, Supplement*, **54**, 259–270

Taylor, B. J. (1988). "The History and Legacy of Photomultiplier Scanners," in Philip, A. G. D., Hayes, D. S., & Adelman, S. J. (eds.), *New Directions in Spectrophotometry*, (Schenectady, NY: L. Davis Press), pp. 3–14

Taylor, B. J. (2007). "The Last Measurements Made with the Wampler Scanner: I. An Analysis of the Consistency and Accuracy of Flux Curves for Bright Standard Stars." *Publications of the Astronomical Society of the Pacific*, **119**, 407–426

Wampler, E.J. (1966). "Scanner Observations of λ4430." *Astrophysical Journal*, **144**, 921–936

Measurement of Polarized Light in Astronomy

Pierre Bastien

1 Introduction

This review deals only with visible and near-infrared (NIR) polarimetry. Stellar magnetic fields are not included, and the choice of topics is necessarily very selective. For more information on most of the topics covered here, and many others, please refer to the proceedings of the conferences *Astronomical Polarimetry: Current Status and Future Directions*, held in Waikoloa, Hawaii in March 2004 and *Astronomical Polarimetry: Science from Small to Large Telescopes*, held in La Malbaie, Québec in July 2008.

2 A Brief History of Polarimetry

The early days of polarimetry are marked by two figures: Bartholin Erasmus (1625–1698), who was a physician and a professor of mathematics at the University of Copenhagen. He is credited for the discovery of double refraction in calcite, or Iceland spar, in 1669. Étienne Malus, a military engineer and captain in Napoléon's army, discovered in 1809 the law now known as the Malus law: $I(\theta) = I(0)\cos^2(\theta)$, which gives the intensity of light after going through two perfect polarizers making an angle θ between them.

2.1 Early Measurements

Soon thereafter, Jean-François Dominique Arago (1786–1853), astronomer, physicist and politician, built the first polarimeter in 1811 and used it for astronomical

P. Bastien (✉)
Département de physique et Centre de recherche en astrophysique du Québec,
Université de Montréal, C.P. 6128, Succ. Centre-ville, Montréal, QC H3C 3J7, Canada
e-mail: bastien@astro.umontreal.ca

E.F. Milone and C. Sterken (eds.), *Astronomical Photometry: Past, Present, and Future*,
Astrophysics and Space Science Library 373, DOI 10.1007/978-1-4419-8050-2_10,

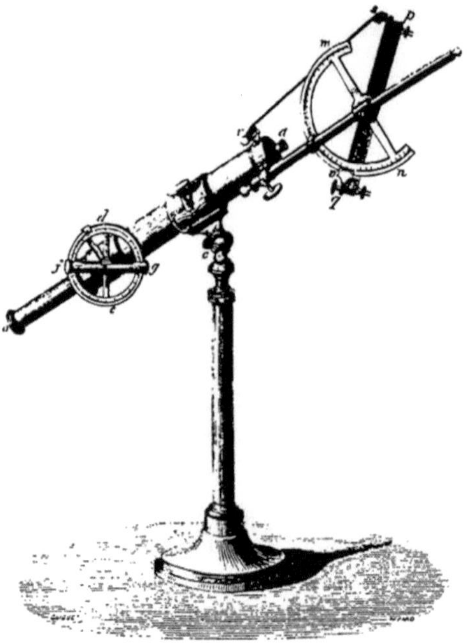

Fig. 1 The first polarimeter ever built. Reprinted from Dougherty and Dollfus (1989), by permission. See text for a description of the instrument

observations (see Fig. 1). Arago used a quartz plate, which acted as a chromatic wave plate such that the plane of vibration of blue light was rotated 130° more than that of red light. The analyzer was a Wollaston prism, which produced two beams. Two images of complementary colors are produced if the light is polarized. We note that the first polarimeter already used the basic principle of all good polarimeters in astronomy today, namely differential photometry![1] Arago's polarimeter was refurbished by Audouin Dollfus around 1986–1988 and is now at the *Observatoire de Paris* museum.

The first extraterrestrial observations by Arago at the *Observatoire de Paris* were aimed at the Moon. He showed that the polarization of solar light reflected by the Moon varies as a function of the lunar phase. The level of polarization was below the detection limit of his polarimeter at Full Moon and maximum around the quarters; it has also a low value near New Moon. Finally, Arago found that the polarization is higher over the seas (maria) than over the continents.

These early lunar polarimetric observations by Arago were confirmed later, see e.g., Dollfus and Bowell (1971) and Fig. 2. We note that a detailed study of lunar polarization was carried out by Dollfus (1955) in his Ph. D. thesis, which was translated later by NASA. It was of great interest in the 1960s when the USA wanted to land the first crew on the Moon, because polarimetry was about the only way to find out about the properties of the lunar surface. Dollfus measured the polarization as

[1] In this case, the differential is in wavelength, not in intensity as is commonly used in current polarimeters.

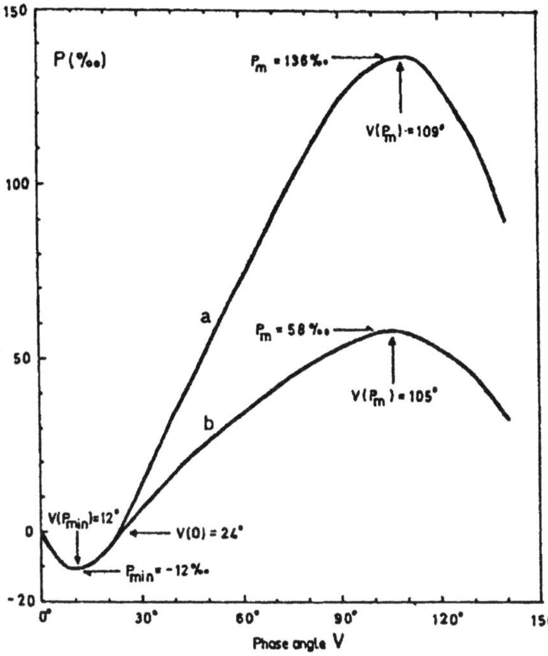

Fig. 2 The polarization of the Moon plotted as a function of the phase angle, the angle between the Sun, the Moon and the observer, so that Full Moon corresponds to 0° and New Moon to 180°. Curve *a* corresponds to polarization measured in Mare Oceanus Procellarum and curve *b* in crater Kruger. The polarization is given in units of permill (=0.1%), a unit no longer used today. Reprinted from Dollfus and Bowell (1971), by permission

a function of the phase angle (=180°− scattering angle). The polarization is negative at small phase angles (less than ≈20°), crosses zero and reaches a maximum around 90–110°. Negative polarization here means that light is polarized parallel to the scattering (or reflection) plane (see (1) below). He found a relation between the slope of the polarization near the inversion point plotted against the albedo, and compared with measurements of samples in the laboratory. The negative and positive branches of the *P*-phase angle curve also contain physical information about the material. Thus he was able to characterize the microstructure or porosity of the lunar surface, hence the interest in the Moon.

Arago (1820) also observed comet C/1819 N1 (Tralles) = 1819 II with his polarimeter and showed that some of the light is polarized due to scattered sunlight. By that time Arago had added a compensating plate (visible in Fig. 1) to his instrument, to make the measurements more quantitative.

After these observations, Arago worked with Fresnel to study the interference of polarized light. Together they formulated the Fresnel–Arago laws (see e.g., Hecht 2001), but they were not able to explain them from basic principles as they assumed that light was a longitudinal phenomenon. They wrestled with this problem for a long time, until Young suggested that light is a transverse wave, i.e., that it vibrates in a direction which is perpendicular to its direction of propagation. Many years later in the context of the Maxwell equations describing electromagnetic properties of light, physicists will state that the electric field associated with a plane wave is perpendicular to its direction of propagation.

2.2 *Towards More Quantitative Measurements*

When making a polarimetric measurement, for example with a polarizer, one measures a vector, by finding the maximum intensity, I_{max}, by rotating the polarizer and noting the corresponding direction, θ. The intensity measured in the perpendicular direction corresponds to a minimum, I_{min}. The polarization is given simply by

$$P = \frac{I_{max} - I_{min}}{I_{max} + I_{min}}. \tag{1}$$

More information about the basics of polarimetry is given in reviews by Bastien (1991), Landstreet (2007), and in these books: Tinbergen (1996), Leroy (2000) and Clarke (2010).

In 1852, George Gabriel Stokes[2] (1819–1903) suggested a new way of representing polarized light using four parameters, which is now named after him, namely the Stokes parameters. However his paper went mostly unnoticed and was discovered much later by Chandrasekhar during his work on the polarization of eclipsing binary stars. It is only since 1960 that its use became more current. The polarization and its direction θ are related to three of the Stokes parameters by the usual equations:

$$P = \frac{\left(Q^2 + U^2\right)^{1/2}}{I}; \quad \theta = \frac{1}{2}\arctan\left(\frac{U}{Q}\right). \tag{2}$$

In 1923 Lyot measured the polarization of Venus and attributed it to scattered sunlight. In 1928, a young Harvard student, Edwin H. Land, invented the polaroid sheet and started a company for manufacturing it, the Polaroid Corporation. Such polarizers are now in great use, in particular in sun glasses.

Three polarization mechanisms will be used in the rest of this paper. When non-spherical grains are aligned by a magnetic field, for example in an interstellar cloud, the grains will polarize a natural or unpolarized light beam which passes through the cloud. The grain absorbs the component of the electric field which corresponds to its most likely orientation, which is with its longest direction perpendicular to the magnetic field. The transmitted component is therefore parallel to the magnetic field direction. This process works particularly well in the ultraviolet, the visible

[2] Stokes is also well known for the Navier–Stokes equations in fluid dynamics and the Stokes theorem.

and the NIR. In the far-IR and in the submillimeter, the grains radiate, and since they are preferentially oriented perpendicular to the magnetic field, the polarization is also perpendicular to the magnetic field. Finally, the grains are very efficient scatterers at wavelengths which are comparable to their size, usually in the UV, visible and NIR, in particular in circumstellar environments. For scattering calculations, Mie theory for spherical grains of arbitrary size is usually used, but more realistic models for, e.g., spheroidal or arbitrarily-shaped grains, are getting more attention lately.

More information about the history of polarimetry can be found in the introduction of the recent book by Clarke (2010).

3 Polarization of the Interstellar Medium

Following the original prediction by Chandrasekhar (1946) that light from eclipsing binaries should be polarized by electron scattering in stellar atmospheres, two observers in the USA tried to verify it. By doing so, Hiltner (1949) and Hall (1949) discovered accidentally that light becomes polarized by dichroic or selective absorption in a medium with aligned dust grains. Following their discovery, Chandrasekhar and Fermi (1953) estimated from observations the magnetic field in the plane of the Galaxy to be about a few microgauss. This also prompted research on grain alignment mechanisms by many theorists since then, starting with Davis and Greenstein (1951)with paramagnetic relaxation experienced by a rotating grain. Today, we believe that dust grains are aligned by radiative torque (see Lazarian (2007) for a review; and an update by Lazarian and Hoang (2011)).

Following the discovery of interstellar polarization, observers obtained polarimetric data on many stars in the Galaxy. It was found that polarization vectors are mostly aligned along the galactic equator, as shown in the map of ≈1,800 stars by Mathewson and Ford (1970).

Probably the next major step was the recognition that interstellar polarization has a particular wavelength dependence, as proposed initially by Serkowski (1973):

$$P(\lambda) = P_{\max} \exp\left[-K(\ln(\lambda/\lambda_{\max})^2)\right], \tag{3}$$

where $P_{\max}$ is the value of the maximum polarization which occurs at the wavelength $\lambda_{\max}$. K is a constant in the original work of Serkowski, but it has later been proposed that it depends also on wavelength.

Lately interstellar polarization is getting more attention because of the quest to measure polarization from the cosmic microwave background. Even thought these measurements are mostly at submillimeter wavelengths, knowing the polarization of the foreground, i.e., interstellar polarization, is critical.

More information about the polarization of the general interstellar medium will be found in the recent review by Bastien (2007). Departures from the usual behavior in the densest regions of molecular clouds are also discussed there.

4 Stellar Polarimetry

A single star is usually not polarized because the scatterers, electrons for hot stars and molecules in the very cool stars, are distributed in a spherically symmetric fashion in their atmospheres such that the total polarization cancels out. In order to get a polarized signal, there must be a break in the symmetry. Possible ways to obtain this asymmetry is to have a flattened star, or an eclipsing binary system, or a binary system with circumstellar scatterers. Also a single star with a non-symmetric distribution of circumstellar scatterers, such as in the case of a young stellar object (YSO) surrounded by a disk and an envelope, can be polarized if viewed from a non-symmetric direction. Some stars, such as Be stars, can have polarization in spectral lines due to an asymmetric extended emission region.

4.1 Binary Stars

The polarization due to electron scattering in envelopes surrounding binary stars has been modelled by Rudy and Kemp (1978) and by Brown et al. (1978), (hereafter BME). If one assumes single scattering and neglects extinction effects in the scattering envelope, then it is possible to integrate over the density distribution of the scatterers and separate out the components into moment integrals. These integrals can be compared directly to a Fourier expansion of the observed values of Q and U into harmonics of $\lambda = 2\pi\phi$, where ϕ is the orbital phase. One can use the coefficients of the first harmonic or those of the second harmonic to compute the inclination of the orbit. Usually the second order coefficients dominate and yield a more precise value of the inclination, as shown in Fig. 3. These coefficients can be used also to compute other parameters of the density distribution, such as its orientation projected on the plane of the sky and many others (see BME and Bastien 1988).

In the case of eclipsing binaries, during the eclipse part of the stellar disk of one star is occulted by its companion, which produces an asymmetric configuration. Chandrasekhar was the first one to compute the polarization of a star at its limb due to electron scattering and predict the effect to be expected during an eclipse. The "Chandrasekhar effect" is illustrated also in Fig. 4. Modeling the eclipse can recover orbital parameters and information about the size of the stars (Fig. 5).

The initial model by BME has been extended by others to consider the error on the inclination angle, the effects of an eccentric orbit, and by Simmons (1983) to consider scattering mechanisms other than Thomson scattering, as long as the scatterers are spherical. To include the extinction from the star to the scatterer and from the scatterer to the observer, one must do the calculations numerically (e.g., Manset and Bastien 2000). More information about polarimetry of binary stars and its modeling can be found in the review by Manset (2005).

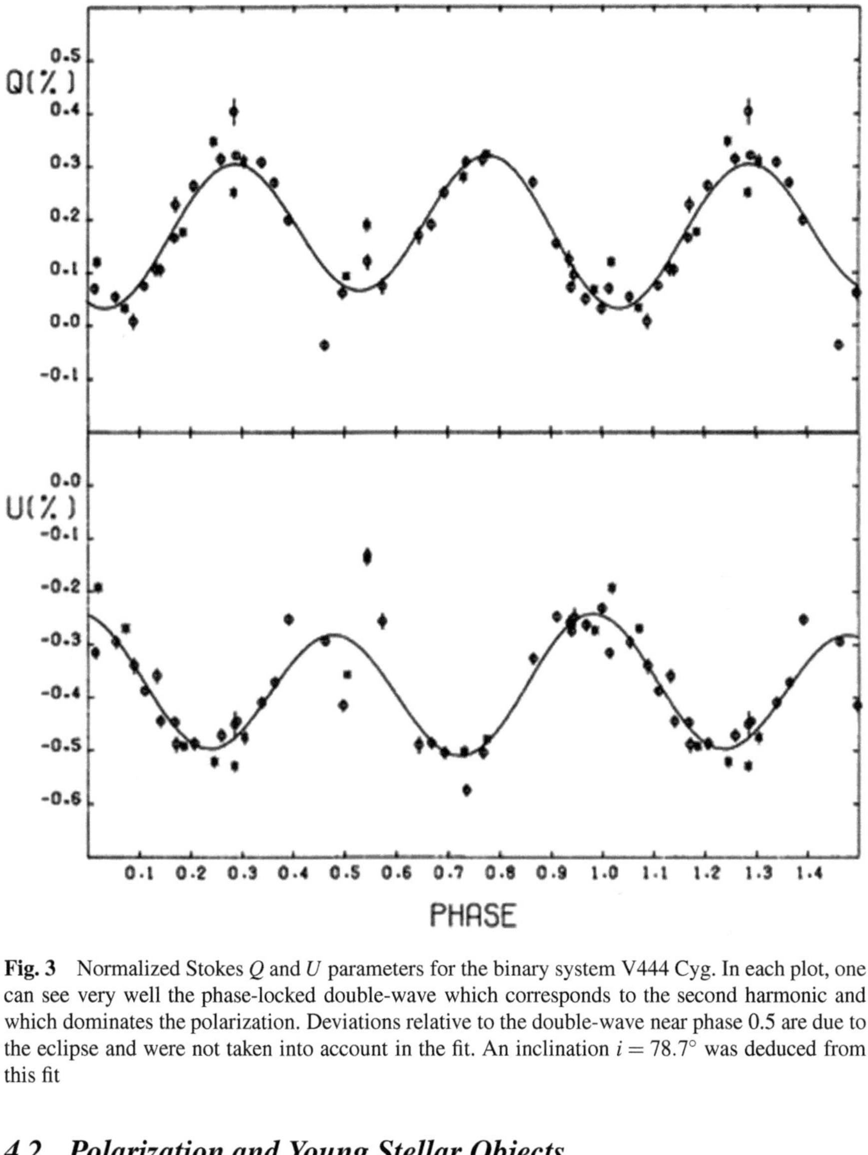

Fig. 3 Normalized Stokes Q and U parameters for the binary system V444 Cyg. In each plot, one can see very well the phase-locked double-wave which corresponds to the second harmonic and which dominates the polarization. Deviations relative to the double-wave near phase 0.5 are due to the eclipse and were not taken into account in the fit. An inclination $i = 78.7°$ was deduced from this fit

4.2 *Polarization and Young Stellar Objects*

Light from Young Stellar Objects (YSOs) is scattered by circumstellar material and produces both linear and circular polarization. The polarization is due to scattering by dust grains in the low-mass T Tauri stars and also by electrons in the more massive Herbig AeBe stars. The circumstellar material comprises in general a disk and an envelope. In the denser regions, multiple scattering occurs. Near the plane of the disk, the polarization maps show a pattern of aligned vectors due to multiple scattering. Further away, in the bipolar cavities where the density is lower due to the outflow, the polarization is usually higher because the photons are scattered only

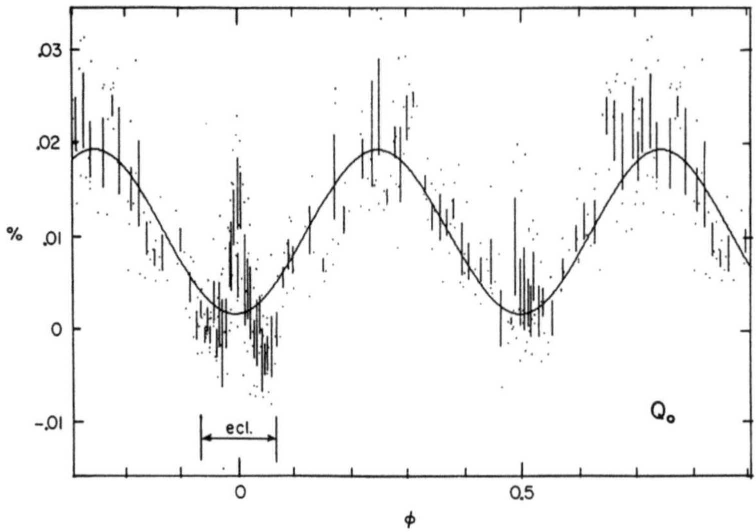

Fig. 4 Normalized Stokes Q parameter plotted as a function of the phase for the eclipsing binary Algol. The double-wave which corresponds to the second harmonic dominates the polarization. Modifications due to the eclipse are visible at phase 0. These measurements by Kemp et al. (1983) were taken in a broad blue-yellow band pass and are among the most precise polarimetric data obtained at that time

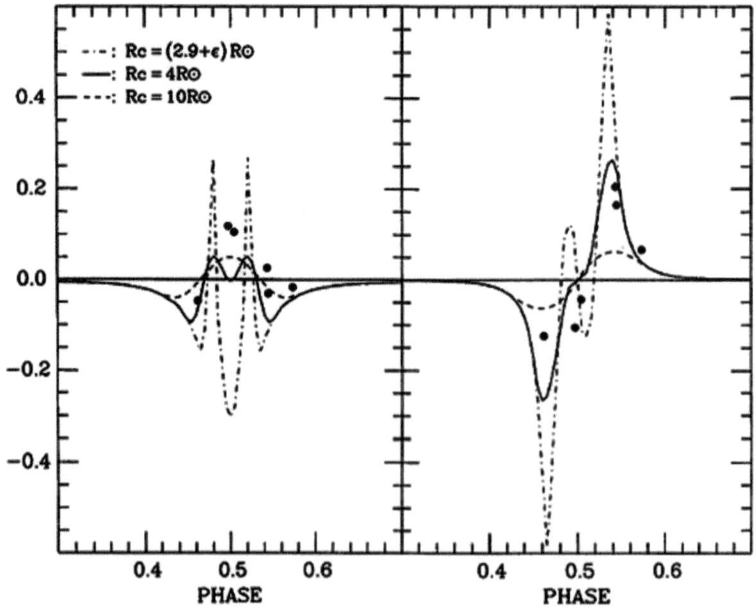

Fig. 5 Normalized Stokes Q and U parameters during the eclipse of the WR + O system V444 Cyg, from Robert et al. (1990). The data points (*filled circles*) are the observed data from Fig. 3 after subtracting off the orbital fit to those values. The *curves* represent a model with different values of the parameter R_c

once. By comparing images and polarization maps with Monte Carlo calculations of the multiple scattering in the circumstellar envelopes, one can get various information such as the inclination of the disk, the size and optical properties of the scattering particles, etc.

Current research topics on YSOs include grain properties, dust stratification in disks as a result of gravitational settling, and dust growth in circumstellar disks. See the excellent review by Watson et al. (2007) for more details.

5 The Present and the Future

Polarimetry is done in photon-counting instruments and also in imaging mode, the latter being used particularly on large telescopes. Here are some typical examples of polarimetry on mid-size to large telescopes. Of course, many other papers could have been selected. Beauty and the Beast (Manset and Bastien 1995), a photon counting polarimeter on the 1.6-m Mont-Mégantic telescope, was used for extensive studies of young stars, e.g., Manset and Bastien (2002, 2003). Imaging polarimetry on the 2-m Bernard-Lyot telescope at Pic du Midi (Monin et al. 1998) and on the 8-m VLT/FORS1 (Monin et al. 2006) was used to study disks alignment in pre-main sequence binary stars.

Imaging polarimetry is also used with adaptive optics systems to increase the spatial resolution: Close et al. (1997) used the University of Hawaii adaptive optics system on the 3.6 m Canada–France–Hawaii telescope to resolve the young binary star R Mon and measure a separation of $0.69''$. Murakawa et al. (2008) used adaptive optics systems on both VLT and Subaru telescopes for a more detailed study of the same star in the NIR.

Another trend in current polarimetry research is the quest for better precision and accuracy. The research on exoplanets has certainly provided a major incentive for achieving higher accuracy from optical polarimeters, since the first evidence for an exoplanet around a main-sequence star, 51 Peg, by Mayor and Queloz (1995) or the first imaging of a planetary system, HR 8799, by Marois et al. (2008).

The instrument PlanetPol has pushed the limit to 1 part in 10^6 (Hough et al. 2006) or about a factor of 100 better than current polarimeters. The major difference with previous polarimeters, which achieve a precision of $\sim 10^{-4}$, is that the first component after the two telescope reflections is a photoelastic modulator which modulates the polarized signal at typically 20 kHz. In order to achieve this precision, extreme care is required, as explained by Hough et al. (1996). Telescope and instrument polarization must be measured and subtracted from the source data.

The light from a planet is polarized due to scattering by molecules and aerosols in its upper atmosphere. The polarization varies around the orbit of the planet as the scattering angle changes. Seager et al. (2000) evaluated the polarization, including the dilution from the mostly unpolarized stellar light, to be a few times 10^{-6}. However, they point out that the predicted magnitudes of the light and polarization curves are highly dependent on the sizes and types of condensates in the planetary atmosphere. If we could separate the light from the planet itself, the polarization would be ∼10%. Lucas et al. (2009) reported very sensitive polarization from the stars with hot Jupiters 55 Cnc and τ Boo. They used the standard deviations of their measured Q/I and U/I to put upper limits on the geometric albedo of the planets, for a given radius. The higher measured standard deviations for τ Boo could be related to spot activity which was detected photometrically by the MOST satellite.

Of course, many other research areas should be revisited with these new very sensitive polarimeters.

Another area where a very high accuracy is required is in the recent quest to measure the polarization from the cosmic microwave background in the submillimeter. In this case, the polarization is due to Thomson scattering of the photon flux with a quadrupolar anisotropy (Aumont 2011). However, the foreground polarization from dust in our Galaxy will have to be removed from the observed data, hence the need to characterize it very well.

Unfortunately, the future very large telescopes will have at least one oblique reflection before the polarization modulator, due to the very large beam after the second mirror. The implications are that absolute polarimetry will be more difficult than with a conventional set up, where the polarimeter is fixed at a Cassegrain focus.

6 Concluding Remarks

Polarimetry has made great progress from new inventions, both instrumental and theoretical. Polarimetry will have been carried out with instruments on telescopes ranging from 3-cm to 30-m (or even 42-m) diameter in slightly more than 200 years, from 1811 to approximately 2020.

Polarimetry is in a sense differential photometry; vectorial information is obtained. It yields information about the geometry of sources, the inclination of stellar or planetary orbits, the size and optical properties of scatterers, the orientation and magnitude of magnetic fields, etc. Polarimetry complements the information gained from photometry and spectroscopy.

Very exciting new developments lie ahead for polarimetry, for example in the areas of exoplanets, active galaxies, the CMB, and other areas, some of them unexpected. Large telescopes will be most useful if polarimetry is taken into account in their initial design, not as an add-on feature to an already existing facility.

Acknowledgments It is a pleasure to thank Eugene Milone for the invitation to the special HAD-AAS session that led, ultimately, to this paper. My work on polarimetry throughout the years has been supported by NSERC grants.

References

Arago, J. F. D. (1820). *Annales de chimie et de physique*, **13**, 104

Aumont, J. (2011). "Astronomical Photometry: Science from Small to Large Telescopes", in P. Bastien & N. Manset (eds.), (San Francisco: ASP) *ASP Conference Series* (in prep.)

Bastien, P. (1988). In G. V. Coyne, et al. (eds.), *Polarized Radiation of Circumstellar Origin*, (Vatican: Vatican Press), p. 595
Bastien, P. (1991) In C. J. Lada & N. D. Kylafis (eds.), *The Physics of Star Formation and Early Stellar Evolution*, (Dordrecht: Kluwer Academic), p. 709
Bastien, P. (2007). "Sky Polarisation at Far-Infrared to Radio Wavelengths: The Galactic Screen before the Cosmic Microwave Background," in M.-A. Miville-Deschênes & F. Boulanger (eds.), *EAS Publication Series* **23**, 131
Brown, J. C., Aspin, C., Simmons, J. F. L., & McLean, I. S. (1978). *Monthly Notices of the Royal Astronomical Society*, **198**, 787
Chandrasekhar, S. S. (1946). *Astrophysical Journal*, **103**, 365
Chandrasekhar, S. S., & Fermi, E. (1953). *Astrophysical Journal*, **118**, 113
Clarke, D. F. (2010). *Stellar Polarimetry*. Berlin: Wiley-VCH
Close, L. M., Roddier, F., Hora, J. L., et al. (1997). *Astrophysical Journal*, **489**, 210
Davis, L., & Greenstein, J. L. (1951). *Astrophysical Journal*, **114**, 206
Dollfus, A. (1955). (Thesis, Université de Paris). *NASA Technical Translation F-188*, 1964
Dollfus, A., & Bowell, E. (1971). *Astronomy and Astrophysics*, **10**, 29
Dougherty, L. M., & Dollfus, A. (1989). *Journal of the British Astronomical Association*, **99**, 183
Hall, J. S. (1949). *Science*, **109**, 166
Hecht, E. (2001). *Optics*, (4th ed.), (Reading: Addison-Wesley), Chap. 9
Hiltner, W. A. (1949). *Science*, **109**, 165; *Astrophysical Journal*, **109**, 471
Hough, J., et al. (1996). *Publications of the Astronomical Society of the Pacific*, **118**, 1302
Hough, J. H., Lucas, P. W., Bailey, J. A., Tamura, M., Hirst, E., Harrison, D., & Bartholomew-Biggs, M. (2006). *Publications of the Astronomical Society of the Pacific*, **118**, 1302
Kemp., J. C., Henson, G. D., Barbour, M. S., Kraus, D. J., & Collins, G. W., II (1983). *Astrophysical Journal*, **273**, L85
Landstreet, J. D. (2007). "The Future of Photometric, Spectrophotometric and Polarimetric Standardization", in C. Sterken (ed.), (San Francisco: ASP) *ASP Conference Series* Vol. **364**, p. 481.
Lazarian, A. (2007). *Journal of Quantitative Spectroscopy and Radiative Transfer*, **106**, 225
Lazarian, A., & Hoang, T. (2011). "Astronomical Polarimetry: Science from Small to Large Telescopes," in P. Bastien & N. Manset (eds.), (San Francisco: ASP) *ASP Conference Series* (in prep.)
Leroy, J.-L. (2000). *Polarization of Light and Astronomical Observations*. (Amsterdam: Gordon and Breach) (also available in French, *La polarisation de la lumière et l'observation astronomique*, 1998)
Lucas, P. W., et al. (2009). *Monthly Notices of the Royal Astronomical Society*, **393**, 229
Manset, N. (2005). "Astronomical Polarimetry: Current Status and Future Directions," in A. A. Adamson et al., (eds.), (San Francisco: ASP) *ASP Conference Series*, Vol. **343**, p. 389
Manset, N., & Bastien, P. (1995). *Publications of the Astronomical Society of the Pacific*, **107**, 483
Manset, N., & Bastien, P. (2000). *Astronomical Journal*, **120**, 413
Manset, N., & Bastien, P. (2002). *Astronomical Journal*, **124**, 1089
Manset, N., & Bastien, P. (2003). *Astronomical Journal*, **125**, 3274
Marois, C., et al. (2008). *Science*, **5906**, 1348
Mathewson, D. S., & Ford, V. L. (1970). *Memoirs of the Royal Astronomical Society*, **74**, 139
Mayor, M., & Queloz, D. (1995). *Nature*, **378**, 355
Monin, J.-L., Ménard, F., & Duchêne, G. (1998). *Astronomy and Astrophysics*, **339**, 113
Monin, J.-L., Ménard, F., & Peretto, N. (2006). *Astronomy and Astrophysics*, **446**, 201
Murakawa, K., Preibisch, T., Kraus, S., et al. (2008). *Astronomy and Astrophysics*, **488**, L75
Robert. C., Moffat, A. F. J., Bastien, P., St-Louis, N., & Drissen, L. (1990). *Astrophysical Journal*, **359**, 211
Rudy, R. J., & Kemp, J. C. (1978). *Astrophysical Journal*, **221**, 200
Seager, S., Whitney, B. A., & Sasselov, D. D. (2000). *Astrophysical Journal*, **540**, 504

Serkowski, K. (1973). "Interstellar Dust and Related Topics," in J. M. Greenberg, & H. C. van de Hulst (eds.), (Dordrecht: Reidel) *IAU Symposium* **52**, p. 145
Simmons, J. F. L. (1983). *Monthly Notices of the Royal Astronomical Society*, **205**, 153
Tinbergen, J. (1996). *Astronomical Polarimetry*. (Cambridge: Cambridge University Press)
Watson, A. M., Stapelfeldt, K. R., Wood, K., & Ménard, F. (2007). In B. Reipurth, D. Jewitt, & K. Keil (eds.), *Protostars and Planets. V.* (Tucson: University of Arizona Press), p. 523

Index

E.F. Milone and C. Sterken (eds.), *Astronomical Photometry: Past, Present, and Future*, Astrophysics and Space Science Library 373, DOI 10.1007/978-1-4419-8050-2,

MIX
Papier aus verantwortungsvollen Quellen
Paper from responsible sources
FSC® C105338

If you have any concerns about our products,
you can contact us on
ProductSafety@springernature.com

In case Publisher is established outside the EU,
the EU authorized representative is:
Springer Nature Customer Service Center GmbH
Europaplatz 3, 69115 Heidelberg, Germany

Printed by Libri Plureos GmbH
in Hamburg, Germany